HIGH NOON FOR NATURAL GAS

The New Energy Crisis

Julian Darley

CHELSEA GREEN PUBLISHING COMPANY
WHITE RIVER JUNCTION, VERMONT

Project Editor: Collette Leonard
Developmental Editor: Catherine Tudish
Copy Editor: Cannon Labrie
Proofreader: Lori Lewis
Indexer: Peggy Holloway
Design and Composition: Terrianne Klein, Sterling Hill Productions
Printed in the United States
First printing, July 2004
10 9 8 7 6 5 4 3 2 1

Printed on acid-free, recycled paper

Library of Congress Cataloging-in-Publication Data

Darley, Julian, 1958-
High noon for natural gas : the new energy crisis / Julian Darley.
   p. cm.
ISBN 1-931498-53-9 (pbk.) – ISBN 1-931498-67-9
1. Natural gas. 2. Fuel switching. 3. Energy policy. 4. Natural
gas–Economic aspects. 5. Natural gas–Environmental aspects. 6. Natural
gas–Government policy. I. Title.
TP350.D37 2004
333.8′233–dc22
                                          2004004708

Chelsea Green Publishing Company
Post Office Box 428
White River Junction, VT 05001
Editorial and Sales offices: (802) 295-6300
To place an order: (800) 639-4099
www.chelseagreen.com

Contents

Acknowledgments

I would like to express my deepest thanks to the following petroleum geologists, who have unstintingly lent me the wealth of their knowledge and helped me personally with tireless patience. I should also say that it has been an utmost pleasure to know such people, and that along the complex path of reaching an understanding of Cinderella gas, they have shared with me many bizarre and amusing stories. Gas as we burn it may be dry, but the geologists who helped find it are anything but! In no particular order, I thank Walter Youngquist, Jean Laherrère, and Colin Campbell. I would also like to thank some non-geologists: Matt Simmons, who first alerted me to the dangerous situation of U.S. natural gas supply, whilst I was pursuing the better known subject of peak oil; Paul Sears, whose understanding of Canadian gas gave extra urgency to my task; Mike Ruppert and Chris Skrebowski, who published my first accounts of U.S. gas problems; and Richard Heinberg, who has both offered wise counsel and been an inspiration with his forthright writing.

Foreword

Natural gas is easy to take for granted: not only is it invisible, but it is delivered to our homes in pipes we seldom see. We don't have to pull into a filling station every week or so to replenish our supply. Indeed, for many of us, our only direct experience of it is the blue flame on our stovetop—though we maintain an indirect connection via our monthly utility bill. We assume that gas will always be there for us—cheap, clean burning, and efficient.

Gas is also increasingly important for electricity generation and for many industrial purposes, including the synthesis of chemical fertilizers. Its many advantages have combined to make it our second most important energy source after oil.

Environmentalists have tended to prefer natural gas to the other fossil fuels because its molecules contain the fewest carbon atoms and thus produce the fewest pollutants (including CO_2) when burned. Indeed, the strategy of many countries for meeting the Kyoto carbon mandates has involved switching from the use of coal to natural gas for power generation. Natural gas is also the cheapest feedstock for producing hydrogen—supposedly the basis for our next energy economy.

Nevertheless, at the end of the day natural gas, like the other fossil fuels, is nonrenewable and thus limited in quantity. As Julian Darley tells us in this important and timely book, the overall picture of gas's future is worrisome. While industry and government officials have led consumers to believe that we can rely on cheap and abundant natural gas for many decades to come, the reality—especially in North America—is that the days of *cheap* gas are already over. North America has passed its historic peak of production, and the United States, the world's foremost user, will have to begin importing more gas from elsewhere. However, in just a couple of decades extraction rates for the world as a whole will start to decline.

Natural gas is a sunset commodity. As such, it constitutes an important chapter in the sobering story of energy-resource depletion— a story that is already playing out toward dangerous and possibly tragic conclusions in the case of oil. As consuming nations like the United States become ever more dependent on imported energy resources, the economic and geopolitical consequences are likely to be dire. The past century has seen numerous oil wars (World War II, the most destructive conflict in history, was in large part fought over access to petroleum reserves). With increasing international trade in diminishing supplies of natural gas, we may begin to see gas wars as well.

My friend and colleague Julian Darley brings exactly the right sensibility to this topic: He has an ecological-systems perspective shaped by many years spent studying population and resources. Moreover, he is a journalist who is passionately interested in identifying important issues and researching them thoroughly, talking extensively to the most knowledgeable sources, making credible and informed assessments, and getting the word out to as broad an audience as possible, even if that means bypassing the usual commercial media channels (hence his pioneering effort in establishing the www.globalpublicmedia.com website).

When, in the summer of 2003, U.S. energy secretary Spencer Abraham decided to convene a blue-ribbon panel to consider the developing North American gas crisis, Julian hopped on a plane to Washington and covered the hearings. When the residents of Vallejo, California, defeated an industry plan to site an LNG terminal near their city, he went to Vallejo and talked to the key players. When petroleum investment banker Matthew Simmons began making alarming public statements about a looming gas crisis that is likely to snowball throughout the remainder of the decade, Darley interviewed Simmons repeatedly.

Meanwhile, the mainstream media continue to treat natural gas as a minor subject to be addressed somewhere in the business pages, if at all. Their reporting is typically inconsistent, fragmentary, and thin on context. Even when a front-page story appears (as when Federal Reserve chairman Greenspan declared, in June 2003, that natural gas shortages in coming years could undermine the U.S. economy), readers of the major dailies were given little background information to help them understand the depth and seriousness of the problem.

Hence the need for this book.

Fossil fuels are the foundation for our modern industrial way of life. Their depletion represents a fundamental challenge to the continued existence of much that we take for granted—well-lit and orderly cities, plentiful consumer goods, cheap and abundant food, instant communications, and fast travel. As first North American and then world extraction of natural gas dwindles, our entire global system of production and distribution will be challenged in unprecedented ways.

Julian Darley conveys a dramatic overview of the situation as few others have done; what is more important, he has given long and careful thought to the vital question, *How shall we deal with this problem?*

As Julian makes clear, a response along the lines of simply searching for substitute energy sources will not be sufficient. We are too many people extracting too many resources too quickly from a finite planet. We must learn to use much less of virtually everything. Yes, we should find alternative energy sources—ones that are renewable and as environmentally benign as possible. But if we assume that we can do so and then continue on as we are now, we are in for a series of rude surprises. Today the declining resources are oil and natural gas; tomorrow they will be freshwater, copper, lead, phosphates, fish, and topsoil. Fossil fuel depletion is merely one of nature's first wake-up calls, letting us know that industrialism as we have been pursuing it is fundamentally unsustainable.

If we are to shake ourselves from slumber and begin to make the needed changes, we first need good information. Our political leaders are incapable of making the hard choices that are desperately needed until the people comprehend the dilemma and demand action. I believe that the people of the world—Americans included—will voluntarily undertake the considerable effort and sacrifice entailed in downsizing, relocalizing, and slowing our industrial way of life *if* they fully understand the consequences of *not* doing so. But they will only achieve that understanding if information such as is contained in this book is widely and quickly distributed.

What we do in the next few years in response to gas and oil depletion will largely determine how the remainder of this century unfolds. It would be a mere platitude to say that much is at stake. In fact, virtually *everything* is at stake—including the survival of our species and of most others. The path we are currently pursuing will entail spiraling global climate change, unending resource wars, and deepening economic chaos. The alternative will require cooperation and self-limitation on a scale never attempted by any human society.

Go ahead. Turn up the burner on your gas range. Meditate on those blue flames and all that they imply. If you have children or grandchildren, think about them and the conditions in which they may be forced to live as the planet's finite hydrocarbon energy resources are burned once and for all, and as competition for what remains intensifies. And think about what you can do to help steer your family, your community, your nation, and the world as a whole toward a survivable transition to a peaceful post-hydrocarbon future.

Richard Heinberg

Preface

The primary audience for this book will be those living in the United States and Canada, countries that are very closely linked—even if their relationship is lopsided. Per capita, both countries consume significant amounts of energy, including natural gas. It is quite certain that what has a negative effect on the United States will not be good for Canada; nor will it be good for the rest of the world because of the dominance of the U.S. economy. Some nations and peoples, such as Britain, already face situations remarkably similar to those of the United States and Canada; many others will soon be adversely affected by the shock waves created as the economic giants try to deal with an unprecedented energy shortage.

Personal Note

I have some critical things to say about the United States, and some of it may sound anti-American. I would therefore like to say the following: Without any doubt the best and most enjoyable years of my life were spent in America, and I look forward with a thrill every time

I go there to visit friends or to give talks and lectures about the seriousness of the energy situation. The kindness and generosity that
Americans have shown me is one of the reasons why I have written this
book. Though I have met many Americans in my life, and know some
of them well, nonetheless it is a tiny and doubtless unrepresentative
fraction. Yet these people suggest that there must be millions of
Americans similarly horrified by what is being done in their name by
their national leaders, and by the corporations that exert far-reaching
control in the United States and beyond. Americans who care know
that their country has descended into something resembling fascism—
the merger of government and corporate power. Unfortunately, it has
been that way for a lot longer than many realize. Strangely enough,
though this book is about an invisible, tasteless, odorless gas, it is part
of the answer to slowing the monster of destruction ravaging our
planet. Energy is one of the keys, perhaps *the* key, and it is one of the
few places where ordinary communities and citizens, people I call the
"walking worried," can act together and do something very real that
will help themselves, help others, and change the structure and system
causing the breakdown of our biosphere.

Dedication

I therefore dedicate this book first to the millions of brave Americans
who don't believe in endless growth, endless corporate power, and endless, illegal, undemocratic government. This book is not an elegy or a
funeral oration; it is a tool with which to take action.

I also dedicate this book to my late mother, whose memory never
leaves me. As a nurse, she spent her life caring for people who were
weak and suffering. She did this unceasingly, whether she was paid or
not. She couldn't help it, she was just made that way. From her example
I learned what I believe to be a central principle of human life: that the

strong shall protect the weak. This is the very opposite of what most governments and all global corporations believe and embody.

I dedicate this book also to all those who have helped me at any time during my life, and especially to my wife, Celine, without whom this book would not have been written, and certainly not finished, and without whom life would be a pretty bleak affair.

1

Introduction

Why "High Noon"?

For those who haven't seen the film *High Noon,* Gary Cooper plays Will Kane, a man unwilling to run away from a nasty set of problems—the worst of whom, Frankie Miller, is due to arrive in town on the noon train. Kane's problems are also the problems of the people in his town, but they refuse to confront the facts, hide in the comfort of their homes, and won't help him deal with reality. They hope that everything will be all right if they leave well enough alone. In the end, one person does help Will, namely, his wife. The cowardice of the townspeople is chilling, as is their willingness to leave to others what they should really do themselves. Unfortunately, being a cowboy film, albeit one of the best, homicide and violence are the methods chosen to settle matters.

In North America, particularly the United States and Canada, there are some energy problems coming to town that are every bit as menacing as Miller's gang. So far, the United States has dealt with the oil problem by using corporate and military force—most dreadfully,

1

with the invasion and occupation of Iraq in 2003. As so often happens with violent solutions, however, things didn't go quite according to plan, and by early 2004, though oil production has risen from immediate post-war levels, operations are still plagued by sabotage and technical problems. Canada's response to U.S. oil supply difficulties was to pump its own easy oil flat out and sell more than half to the United States. But with the easy oil in decline, Canada and the United States believe that the Albertan tar sands will be their savior. That belief may not be well founded, but the reasons why go far beyond the cold Canadian tar sands and will soon touch the whole world.

The U.S. energy story is complex and not very edifying in places, but it is full of wild and unexpected twists. The strangest is that the worst immediate problem confronting the United States (and Canada) is not oil, but natural gas. It is a natural gas shortage that could seriously interrupt the U.S. economy, though there are several other impressive candidates, including a dollar crash and global oil peak. ("Peak" refers to a peak in extraction, followed by inexorable decline. Peak production can apply to the extraction of any naturally occurring fluid or gas deposit).

The information to support the case being made here is largely in the public domain, but few have noticed. In September 2003, the National Petroleum Council (NPC), a group of oil and gas industry advisers, released their eighteen-month-long study of the state of North American gas supply and demand. It should have been at the top of the best-seller list. As it was, just a handful of energy analysts reviewed it, and they are calling out loud and clear that the Titanic is about to hit the iceberg, but, if you will pardon the mixture of metaphors, mission control is not listening.

The coming shortage of natural gas in the United States and Canada, compounded by global oil peak and decline, will try the energy and economic systems of both countries to their limits. It will plunge first the United States, then Canada, into a carbon chasm, a hydrocarbon hole, from which they will be hard put to emerge unscathed.

A Great Triumph

The United States appeared to make it through the 1970s oil shocks not only unshaken but immensely strengthened. Later, its great theoretical enemy, the Soviet Union, lay vanquished after the collapse of Communism. The United States had won, and the twenty-first century would now certainly be America's for the taking. However, despite appearances, and triumphalist rhetoric to the contrary, the United States was in reality grievously wounded by the oil shocks. Even so, after 1980, and further oil price increases following the 1979 Iranian revolution, instead of instituting a program of even more vigorous energy constraints, the United States embarked on a tide of unparalleled waste and consumption, so that twenty-five years later, it now imports more oil than at any time in its history. But at least it can import that oil easily. Natural gas cannot be imported on demand; it requires planning long in advance, especially if it comes over the ocean as liquefied natural gas, or LNG. Having ignored all the warnings and squandered two decades of planning time, the United States is left with gas from across the oceans as the only possible business-as-usual option. But it is too late.

Two well-respected energy experts, the world's foremost private energy banker, Matt Simmons, and energy analyst, Andrew Weissman, have been warning with great conviction and evidence that the system is heading for disaster. They have both said that North American gas supply numbers show that economic growth in America is under severe threat. Simmons went further to say that, compounded by the global peak in oil production, the world economy will be badly constrained, and once the full force of oil decline begins to set in, growth as we know it will be over. These men are certainly not anticapitalist radicals, indeed one of them has been an adviser to President George W. Bush. Although the petroleum industry has a history of creating scares to drive up the price of oil and gas, such tactics are unwise when the supply is constrained by geology.

For the United States, and by extension for Canada and Mexico, high noon has indeed arrived for natural gas. Why should the rest of the world care? The United States, at least, is not a popular nation in the world community at present. Its seemingly unstoppable violence is visited on friend and foe alike, as the British, Canadians, Hungarians, Afghanis, Iraqis, and countless others have witnessed. All this violence for what? Once there were ideological overlays, however specious, but now it is becoming clear that American aggression is in great part about energy, once mainly oil, but increasingly natural gas as well. Even if the world doesn't want to care much about all the human suffering caused by the global war for energy, individual nations should worry that they may be the next target if they have oil and gas, and if not, that they will be competing with the United States for the limited supply of fossil hydrocarbons, the stuff that makes the modern industrial world work.

For North America, now racing full bore into the carbon chasm (a huge gap between high energy demand and falling supply), the question boils down to whether or not to agree to build dozens of new liquefied natural gas importing terminals. This will force the United States to become ever more dependent on foreign and mainly hostile nations for an energy source that supplies a quarter of its needs. Canada and Mexico will be faced with similar dilemmas, though less open to hostility.

Clear Warning?

The rest of the world is not far behind. There is still a degree of choice left for some about whether to make the same mistake as North America (and Britain) and switch en masse to the last great hydrocarbon, natural gas. If this is the chosen path, it will lead inexorably, as global natural gas peaks and declines, to a future dominated by nuclear power and coal, which will even more surely condemn our planet to global warming and grim battles for the remaining oil and gas that could all too easily trigger global warfare.

Or else we could choose, at last, to try to live within our daily ration of energy supplied by the sun, however ludicrous this may sound. We could choose to end the cycle of violence that has characterized much of recorded human history and that has reached an astonishing crescendo of destruction in the last one hundred years. These energy questions are some of the starkest and most historic ever placed before the human race. However we choose to answer the energy question, the problem of human overpopulation, so long a taboo subject for almost everyone, will resurface because of the coming shortages of natural gas and oil. This time we will not be able to smother the issue with fossil-fueled food.

The supply constraint of natural gas is so serious and coming so soon in North America that whichever way the LNG situation is resolved, Americans, Canadians, and Mexicans will be faced with using less natural gas. This is not a rerun of the 1973 energy shortages; it is much more serious and permanent. If by a miracle of heroism and fortitude, Americans decide not to go for acquiring massive new supplies of LNG, not to "globalize gas" as market proponents describe it, then other grave questions will have to be faced. In fact, they will have to be faced anyway, but this will be the last time, because after natural gas, there are no more easy energy bonanzas left in our planet's crust, and now the first indisputable warnings about the limits to global natural gas reserves have appeared. In both 2001 and 2002, for the first time ever, more natural gas was used than found. Exactly twenty years ago, the same thing happened with oil: more was used than found. It wasn't a blip, it was a permanent trend. The present century has seen very little oil discovered, and 2003 was the worst year for vital, large discoveries in many decades. Now the same thing is happening to natural gas. It is the dread hand of depletion writing on the wall; we can ignore it for sure, but it will not ignore us.

A Global Problem

There are many other questions raised by natural gas, this invisible fuel and feedstock. Some of them are quite bizarre, many are curious and surprising. But none of them are nice surprises. Although each country will face a different set of questions, there is nowhere on Earth that will not be affected in some way by the issues raised here. Like it or not, personal and policy decisions will have to be made everywhere in regard to natural gas. Although it is late, now would be a good time to learn something about natural gas, to understand why we should all start caring about an energy source too long taken for granted.

Much of this book concerns the nature of natural gas, its disposition, its supply, and why we demand so much of it. But though the book is about natural gas, it is also *not* about natural gas. Rather, it is about energy and the future of humanity. It is about life itself, and some of the limits to the systems that support our industrial way of life. The final chapter will explore why economic growth, one of the strongest drivers of increased gas consumption, is forced upon us. I shall also discuss how the public—given that our institutions have mostly failed us—can meet the extraordinary challenge of using both less natural gas and less energy in general. Many people do accept that we live in a limited world and want to learn to live in it without damaging it any further, but we shall have to rebuild those broken institutions and infrastructures first. Some, indeed, may have to be invented for the first time.

This will be a journey, at times complex but also intriguing and strange. It will be necessary to pick a way through a jungle of confusing terms and numbers, so that we can begin to understand what the global petroleum corporations, and many of their client governments, do not want to us to understand or ask any awkward questions about.

Above all, the fate of Gary Cooper in *High Noon* should be avoided.

He couldn't persuade the public to help him, and he had to resort to lone violence. In the end, in his view, he had no choice. All too often, if we leave a situation until too late, the best options may be closed.

Energy Literacy

Because energy is personal and political, awkward and unpleasant questions quickly arise. Energy is also technical and dangerous, essential both to life and to large-scale, mechanized violence. The discovery of abundant and easy energy in the forms of oil and natural gas has allowed us to disconnect from reality, from realizing that there is only a certain amount of useable energy delivered to the planet every day. We can cheat and run up environmental debts and eat into Earth's fossil energy store as we become ever more addicted to the big energy fix, but as with any debt, any deception, any addiction, there are high costs involved, and these costs are now undeniable.

The situation now arising requires a very deep level of rethinking. One of the requirements is energy literacy, a form of critical thinking and structural rethinking. Energy literacy will require us to notice how much and what type of energy a process of life requires, and where that energy comes from. In the industrial world, we are not used to thinking like this. We have endless tools for thinking superficially about economic goods, but few in policy or practice, for understanding the role of energy, or how it underpins the structures of industrial, automated life. Most importantly, it is not markets, but the mechanisms of our infrastructure that must be understood, and this understanding will be our key critical thinking tool.

Some of the observations and interpretations offered here will make uncomfortable reading, but unfortunately, there is so little structural or critical thinking at any level of society, business, or governance, that neither the tools nor the language for it are well developed or widely practiced. In fact, the opposite is generally true: Critical thinking is actively discouraged by deflection, diversion, and distraction.

Natural Gas Dependency

Perhaps because natural gas is not something that most people have to pump themselves, unlike gasoline, it may seem as magical as electricity. It's always there when you switch it on. However, natural gas is a substance like oil that must be extracted from the earth. Whatever our perceptions of natural gas, it is already the second most important energy source after oil in North America, and will soon be in the rest of the world. If we carry on with the present course, and if the peak and decline of global oil follows some of the more pessimistic predictions, then gas consumption will overtake oil—briefly—until it too declines. In the United States, natural gas is used to make about 20 percent of all electricity. In many places, such as Britain, the percentage is much higher. In the future, the U.S. government, and governments the world over, expect gas usage to rise dramatically. During the last four years, almost 95 percent of the new electricity-generating plants built in the United States have been fueled by natural gas, which means that either gas usage will rise to meet the demand or many operators will go bankrupt when the supply becomes too costly. Canada, as the world's largest producer of hydroelectricity, is not as dependent on natural gas as the United States, but some provinces, especially Ontario, the engine room of the Canadian economy, have less hydroelectric capacity and are therefore more reliant on gas to produce electricity.[1]

While natural gas is becoming ever more important for electrical power production, it is absolutely indispensable across the world for the production of industrial fertilizer. The explosive and highly problematic growth of the human population throughout the twentieth century and into the twenty-first is due in great part to the use of industrial fertilizer. One of the most vital, and presently irreplaceable, feedstocks of that fertilizer is natural gas, which makes natural gas possibly the single most critical ingredient in the diet of many human beings. For a few more years—no one can be sure how long—global natural gas production can increase, but in North America, where gas production has peaked and is declining, food production is already being affected, because of higher fertilizer prices; global oil peak will soon add to the problem. At the same time, water and soil problems are having a negative impact on global food production, especially in China.

There is very little evidence that we can feed six billion or more people without petrochemicals. If water, soil, and climate problems haven't already begun to limit maximum global food production, which does now appear to be the case, once global oil then natural gas supplies become constrained, there will no longer be any way of hiding from our disastrous population explosion. World population before the onslaught of industrial agriculture was around one and a half billion people. We may not even be able to support that number now, without oil and gas, because the very use of the chemicals derived from them has had so many damaging effects on the natural productive capacity of the soil and its related systems.

In the industrial world, natural gas is also critical for heating the places where we live and work. More than half of the United States heats with gas, as does three-quarters of Canada. The percentages in many places in Europe are similar or higher.

Natural gas has numerous other important uses. It is used to make many plastics, fabric for clothing, the injection moldings used everywhere, a huge amount of packaging, and even humble plastic bags. It supplies the heat for many processes from cement production to grain drying and is currently indispensable for making synthetic oil from the Canadian tar sands. Much of the propane and butane used to heat buildings across the world comes from natural gas. Propane is also used for industrial vehicles and smaller trucks, and many cities are now switching their bus fleets to run on natural gas, considering this a wise decision to help reduce global warming. Methanol, useful as a chemical feedstock and for hydrogen fuel cells, is made from natural gas. Almost all hydrogen is made directly from natural gas. A large amount of natural gas is used in oil refining.

The example of methanol highlights the complexity and strangeness of the world of energy and natural gas. What has methanol to do with hydrogen fuel cells, and if hydrogen can be made directly from natural gas, why not do just that? Methanol and hydrogen can be made in ways other than from natural gas (but they are much more expensive). Surely science and economics have taught us that there are always substitutes, including for natural gas? The answer is a resounding yes and no.

Natural gas has become crucial to our continued well-being and will become more so with each passing year. The major oil companies have just begun to say publicly that they are switching away from oil to natural gas as the new primary fossil fuel. It isn't because they want to; they are driven by another ominous factor: the switch to gas is an admission of global oil peak.

Invisibility

To find that we are so dependent on something that we were hardly aware of is surely troubling. If concern moves beyond passive wondering, the question will arise: How did this happen? It has already been pointed out that although natural gas is extremely useful, we do not normally have to handle it ourselves. It is quite literally invisible. Once purified, natural gas has no smell or taste. "Natural" gas seems to be just that, natural, and perhaps endless, like nature itself was once thought to be.

Whatever the rationale, most people in the industrialized world now take natural gas for granted, just as we do electricity, running water, the simple "disappearance" of our human wastes down a white porcelain bowl, and the collection of our garbage and rubbish. This is all natural, we assume, and as things should be. However, these things are not natural, and we are starting to notice that in fact they take a lot of effort, organizing, and energy to achieve. Unfortunately, about the only indicator the public has of problems, or anything else, is price. As we shall see later, price is not a very good indicator for many things, especially natural gas, which is now becoming an endangered species in North America, though not yet in the rest of the world.

Because natural gas is just that, a gas, it is difficult to transport. This is why we need such a complex system of pipelines in North America to deliver it. Until the advent of liquefied natural gas (LNG), it was considered a regional resource and could not be traded over oceans. Natural gas pipelines also crisscross Europe and every country that uses gas in quantity.

Denial

Until very recently, North American natural gas producers have steadfastly refused to admit that they have a problem with supply. Suddenly, in the middle of 2003, amid turmoil in the fertilizer industry and alarm in the chemical and power industries, the producers changed their message and said that indeed supply is in decline. But there is a reason for this public admission, and it is part of the strategy for getting access to what remaining public land and waters are not already being drilled.

The speed and suddenness with which the natural gas crisis appears to have arrived in North America is due in large part to the silence of the extraction industry, but it is also abetted by the myth of inexhaustibility and substitutability of resources. This myth is not confined to America. For instance, many Chinese believe that the vastness of their country, similar in size to the United States or Canada, means endless resources. In the case of oil and gas, this is no longer true for any of these large countries.

In the United States, the boundless-resource myth is part of the religion of progress, the pioneer spirit, and the doctrine of Manifest Destiny. The United States began with some of the largest and easiest-to-access oil and natural gas reserves in the world, and it still has the largest coal reserves on the planet. In the simplest terms, Manifest Destiny means that America is fated to expand forever and has a right to do so, just as the "tree [has a right] to the space of air and the earth suitable for the full expansion of its principle and destiny of growth."[2]

When earlier Americans reached the western limits at the Pacific Ocean, they were still able to buy Alaska. Following the British example in the early nineteenth century, America used the power of capitalism to expand its borders by the proxy of trading and creating markets in other countries, and eventually most of the world. Thus for a long time there did indeed appear to be no limit to resources, or anything else, since most commodities can easily be carried by ship or aircraft, which allows the illusion to continue. Except that gases don't follow the normal shipping rules.

Carbon Chasm

Helped by good fortune, false beliefs, and gruesome myths of justification, most North Americans, whether in positions of power or not, have no idea that there is a serious energy problem at their door. Britain and Ireland are not far behind, and neither is China or New Zealand. Sooner rather than later, all the world's most industrialized and fastest industrializing countries will enter the carbon chasm, a gap that will grow, between increasing fossil energy demand and reducing supply. Although oil will decline globally before gas, because of the transport difficulties, it will be natural gas shortages that hit North America first and hardest.

Yet it was not supposed to be this way. The plan of industry and even many green groups has been to move away from coal and oil to natural gas. Gas is said to be "clean" burning and less harmful for global warming than oil or coal. Natural gas is certainly the last great hydrocarbon on Earth occurring in huge, industrial-scale quantities, accessible with known technology at reasonable cost. Now, both Europeans and Americans are starting to invest huge sums in shipping and constructing pipelines for natural gas, yet few have closely examined whether there is enough gas to justify that investment, let alone whether it is really such a good idea.

The United States and Canada are entering a natural gas crisis, which is becoming evident in 2004 as prices remain high and supply declines. Mexico will be less affected because it is less prone to very cold winters. The United States has been close to a natural gas crisis several times already. Early in 2003, thanks to low levels of natural gas storage, the United States came within days of blackouts and home heating shutdowns. North American supply is simply no longer able to meet desired consumption. Though in Canada there is no outward sign that the problem has been noticed at the ministerial level, the White House has become worried that natural gas will affect the 2004 presidential elections. Some U.S. economists are also worried, and some of U.S. industry is worried, but few North American environmentalists in 2003 were aware that their favored fuel may be in trouble.

If the weather in the United States hadn't been relatively mild in 2002 and 2003, then a full-scale natural gas crisis would have already happened. The fortuitous mild weather allowed the U.S. government some time to stave off such a politically devastating occurrence, relaxing air pollution restrictions in the summer and fall of 2003, so that coal could supplant natural gas in electricity generation, by some estimates saving more than 5 percent of gas consumption. The U.S. Energy Bill, HR 6,[3] which may be passed in some form in 2004, is most likely to allow even more air and water pollution by power companies. Such measures have helped avoid a major crisis in the winter of 2003–04. However, another worrying sign that energy supplies are tight is that North America saw three massive power blackouts in the summer of 2003, the first and largest of which, on August 14, was officially blamed on mistakes made by one power station.[4] Pinning the blame on an individual—corporation or person—helps avoid asking serious questions about the whole system and its overall direction.

The state of natural gas supply is a looming menace, which helps make the whole power grid run increasingly close to the edge. The United States is now turning to more extreme methods to obtain natural gas, including pushing for drilling into more environmentally sensitive areas with more damaging techniques, such as coalbed methane.[5] There are also U.S. plans to build more than two dozen new liquefied natural gas terminals to import gas from the rest of the world, whose gas supply may still be increasing. Both of these actions have enormous and unfortunate ramifications. They also mirror what happened after 1970, when U.S. oil production peaked. Shortly after that came the first great oil shock in 1973, when Persian Gulf oil exporters interrupted supply to the United States and other nations helping Israel fight against Egypt and Syria. It is much easier to disrupt gas supply than oil, a fact that will not be lost on America's foreign enemies.

. For North Americans, the reasons for concern about natural gas are becoming ever clearer. For the rest of the world, there are at least four serious reasons to share that concern. First, whatever America does usually has profound effects around the world. Second, transporting natural gas across oceans entails very expensive decisions with many implications, including military ones, for the producing countries, as well as the consuming nations. Third, both the industrial and less-industrialized nations are starting to convert many major systems to natural gas, especially electrical power stations. This is a mistake of historic proportions, since global gas supply will suffer the same fate as oil. Well before that happens, many other countries will experience the same kinds of natural gas shortages as North America. Citizens will likely find their governments similarly unprepared, even as the gas exploration industry knows that we are already using more gas than we find.

Preparing for High Noon

Now that it is high noon for natural gas, and gas depletion is beginning to bite in the world's largest economy, this book will try to prepare the engaged citizen and the policy maker for what is to come—not just in the United States and Canada, but anywhere in the world.

First the important facts and nature of gas will be described in chapters 2 and 3. Why we use so much gas will be covered in chapter 4, and where the world's gas reserves are located is the subject of chapter 5. Chapters 6, 7, and 8 present an analysis of the political, security, and policy implications of declining gas supply in North America.

The final chapter offers an outline of what ultimately drives gas demand, and what measures, large and small, that we, the public, can take on our own and the planet's behalf.

2

The Gas Itself

What is Natural Gas?

Almost everybody has some idea about oil, and why it is so important, even if the extent of its importance is greatly underestimated. But natural gas is still rather mysterious. Even while using it to heat their houses and cook their dinners, few people really know what it is composed of and how it gets to them.

"Natural" in this context means generated in the earth (or by living things), rather than created by technological means. This was originally to distinguish it from "town gas" or coal gas, which is quite different, being manufactured from coal. "Gas" describes a fluid that has "neither independent shape nor volume" and can expand indefinitely. The term was coined in the early seventeenth century, perhaps 1609, by Jan B. van Helmont, a Belgian scientist working on the composition of the atmosphere.[1] The word itself may be related to both the Greek word *chaos*, which describes what came before matter solidified into the world we know, or the Dutch word *geest*, meaning "spirit."

The Heart of the Matter: Methane

Methane is the lightest and simplest hydrocarbon. Natural gas is composed almost completely of methane, a gas whose molecules contain four hydrogen atoms attached to one carbon atom, making the formula CH_4. Natural gas has a density roughly half that of air, which means that it will rise and disperse when released into the atmosphere. This has certain benefits. It will not sink and collect as some gases do, like carbon monoxide, which is highly toxic and can easily cause explosions. But being lighter than air also has some important and unfortunate consequences, especially since methane is a harmful greenhouse gas. It means that any accidental leakage is hard to restrict. In its pure form, methane is nontoxic to humans. Natural gas also contains small amounts of other useful hydrocarbons such as ethane, butane, propane, and pentane but often also has significant and dangerous amounts of sulfur, mainly in the toxic form of hydrogen sulfide. The presence of sulfur and hydrogen sulfide makes natural gas highly toxic. It also makes production more expensive, since the sulfur must be removed before the natural gas can be sold. Though methane is not toxic in itself, it is certainly highly flammable, and in confined spaces with the right mixture of air, it will and does explode. Many thousands of people have been killed by gas explosions, both in coal mines, factories, and ordinary homes. Being a gas and explosive makes it a very difficult and potentially dangerous commodity to transport.

The Conventions of Gas: Wet, Dry, Stranded, and "Nonassociated"

In the industry, gas as produced and delivered is frequently described as being wet or dry. "Wet" means that the other hydrocarbons that often make up natural gas (in addition to methane), are still present in the product. The gas is "wet" because these other hydrocarbons, while gaseous during extraction, condense at normal or surface temperature, so that once released from the gas field they become liquid. Natural gas may also include water vapor, which adds to the wetness. "Dry" gas means that all the possible liquids, or gases that will easily condense to become liquids, have been removed before delivery.

The other main distinction for gas is whether it is associated or nonassociated. The gas found above and dissolved in many oil deposits is called associated gas. This gas may come out along with oil, whether it is wanted or not. It is often reinjected back into the reservoir to keep the pressure up and help the oil flow up the well. Nonassociated gas means that the field contains mainly natural gas. It is sometimes also called unassociated gas.

Stranded gas refers to natural gas, either associated or nonassociated, that is not connected to a pipeline, or any kind of collection system, usually because the gas deposit is too far from the user. In the past, stranded gas, when produced with oil, was either vented into the air, burned—which the industry calls flaring—or reinjected into the field, usually to keep the pressure up on any oil associated with it, which is what the producers would normally be interested in. Increasingly, as oil gets scarcer and gas becomes the hydrocarbon of choice or last resort, and thus gets more expensive and interesting, there is by definition less and less stranded gas, since pipelines are being built at a great rate to bring stranded gas to market.

Gas reserves are generally, though not very accurately, divided into conventional and unconventional deposits. Conventional gas is that which can be produced at or close to today's prices and with the technology of the day. Thus, what is termed conventional depends on both economics and technology and is constantly changing, though not generally very dramatically. As the price rises, and technology improves, gas that was formerly considered unconventional becomes conventional.

There is no accepted single definition of unconventional gas, but generally it includes gas from tight formations—meaning gas contained in not very porous rocks, shales, and coal seams, also known as coalbed methane, or CBM. Some people also include deep-sea or deep-water gas and Arctic or polar gas in the definition of unconventional. From a production perspective, the most important thing about unconventional gas is that it is either slower or more expensive[2] to extract, or sometimes both. As technology improves, some of these factors may be ameliorated, but the rate of extraction can rarely be increased past a certain point, usually nowhere near the production rate of conventional gas.[3]

The Energy of Gas

Natural gas, like any hydrocarbon, contains a lot of energy. One cubic foot of gas will provide roughly 1,000 Btu, or British thermal units. A Btu is defined as enough energy to raise the temperature of one pound of water by one degree Fahrenheit. A cubic foot of gas will raise half a ton of water by one degree or boil about a gallon. A large, modern, gas-fired power station of 500-megawatts capacity may burn as much as 80 million cubic feet[4] of gas a day.

How Gas Got There

Before you can burn or process natural gas, you have to find it, then get it to where you want it.[5]

Natural gas and oil are often called fossil fuels, because they began as source rock called kerogen ("born of wax"), which is the dead remains of plants, creatures, and microorganisms that lived millions of years ago. Natural gas can be formed either directly from kerogen, or indirectly from oil, in which case gas and oil can be found together.

Although there are a number of different theories about the origins of fossil fuels, it is now generally accepted that the large reservoirs of natural gas (and all oil) were formed by a complex process of several stages. The chief factor in deciding whether oil or gas will be produced is the type of originating kerogen, either sapropel (liptinite) or vitrinite.[6] Oil is formed from algae-rich sapropel (literally, "rotten mud"), whereas gas is formed from vitrinite, made of vegetal material, such as leaves and plants, particularly in large river deltas. Kerogen may be a mixture of both sapropel and vitrinite.[7]

As the organic remains became gradually covered with silt, sand, mud, or more organic matter, to the depth of even a few yards below the seabed, the formation of kerogen began. Slow but powerful geological and tectonic forces then moved the kerogen deeper into the crust. The main factor for such movement is subsidence,[8] which helped form the great sedimentary basins in which so much oil and gas is found.

Once deeply buried, the kerogen, which consists largely of carbohydrates, was compressed by the weight on top of it and intermittently heated by the geothermal gradient[9] to reach temperatures sometimes ranging beyond 500°F (260°C). In this "hydrocarbon kitchen,"[10] over a very long time, usually many millions of years, the episodic conditions of high pressure and considerable temperatures broke some of the carbon bonds in the kerogen, releasing the oxygen, and leaving hydrocarbons in various chemical shapes and sizes. In the case of sapropel source rock, oil generally started to form over about 150°F (65°C), most typically in the range of 170°F (75°C) to 250°F (125°C). Above this temperature, the oil started to break down into natural gas. Since greater depth normally means greater temperature, gas was physically formed below oil, most typically at temperatures up to 350°F (180°C). If the heat rose beyond 500°F (260°C), the hydrocarbon deposits were, or indeed could still be, destroyed by carbonization. Vitrinite, which may be present with sapropel, follows a similar chemical transformation but results mainly in gas.

The range of time involved in gas and oil formation is quite extraordinary. The most ancient known oil or gas field is in eastern Siberia, at Yuruchbeno, though it was only discovered in 1983. It is just over 1 billion years old.[11] From the Precambrian era, there are some source rocks that date back 600 million years, while others were formed as recently as a million years ago, in our own geological period, known as the Quaternary. However, most of the oil and gas that we have extracted in the last 150 years was formed somewhere between 10 million years ago (in the Tertiary period) and 270 million years ago (in the Permian period). In between these two periods came the famous "dinosaur" periods of the Cretaceous, Jurassic, and Triassic.[12] It is because so much of our present industrial system is based on fuel from these periods that we often think of our energy as coming from the age of the dinosaurs. This also gives ready images to those who think we are heading for a similar fate as those once prolific animals.

When natural gas is produced by the action of burial, compression, and heating over vast expanses of time, it is called thermogenic methane, meaning literally "born of heat." Because natural gas is produced at greater depth than oil, unless it has migrated upward, recovering it entails drilling to greater depths than required for oil. This also means that deep hydrocarbons, at 15,000-foot depths or greater, are more likely to be gas than oil. Very deep deposits may be entirely methane. Both because of indirect formation from oil and the more widespread source rock of vitrinite, gas can be formed by a much wider set of geological conditions than oil, with the world's great delta systems, such as the Amazon and the Nile, being particularly favored. On this basis there would be much more gas than oil. However, most gas has leaked away over time.

Unlike oil, natural gas can also be generated by the breaking down of organic material by tiny microorganisms called methanogens. When this is the case, the result is called biogenic methane. Where thermogenic methane is formed deep in the earth, biogenic methane generally occurs near the surface,[13] though it requires the absence of oxygen, otherwise oxidation will take place before any methane is formed. This biogenic process also occurs in the intestines of land animals, including humans. Not surprisingly, but rather unfortunately, most biogenic methane is lost to the atmosphere. There are methods to trap it, however, including the methane that is generated in landfills, which contain large amounts of food thrown away by humans. Food materials could and should be separated, and either composted for use as land fertilizer or collected into larger devices for methane generation.

There may also be "abiogenic" methane, which is reputed to be formed from hydrogen-rich gases and carbon molecules extremely deep under the earth's crust. It is suggested that as these gases rise through the rocks they may react with surrounding minerals, and, in the absence of oxygen and at great pressure, they may form methane deposits.[14] There is considerable skepticism about the existence of this process, however, and the petroleum industry has not shown great zeal in pursuing the idea.[15]

Once created, natural gas, being gaseous and thus much lighter than the solid or liquid material that surrounds it, will try to rise through looser and more porous rocks, especially shales, until it either reaches the surface and escapes into the air, or else is trapped in some way. Natural gas can be trapped by many different geological rock formations, though few are complete seals.[16] Some gas fields are constantly charged or filled by gas coming from the source rocks below.[17] The gas itself will be contained in porous sedimentary rocks, especially those that are clastic, meaning that they consist of broken rocks. Sandstone is one of the most typical reservoir rock formations. Limestones and dolomites are also common.

Above, and possibly around this reservoir, there must be a layer of rock that seals the "leaky," porous rock below. This sealing rock may also be sedimentary, but will not be like sandstone. It could be shale, but salt (and anhydrite[18]) is the most effective. In fact, in many reservoirs, the gas is on top of oil, which itself is above water. This latter may seem strange, but oil is lighter than water,[19] which is why oil spills at sea are so disastrous, since the oil floats and then sticks to whatever solid thing gets in its way, including birds, reefs, and marine life. At normal temperatures, natural gas, being lighter than both water and air, cannot spill but will rise upward. Unfortunately, for the climate, this could be even more problematic than an oil spill, since methane is a greenhouse gas more than twenty times as damaging as carbon dioxide.

If the sealing system is effective, however, the gas will remain trapped until there is some geological or seismic event (such as erosion or a fault), or until a drill bit finds it.

Conventional Gas

As the world moves into the inevitable oil peak and decline, there is, at least in the business press, more notice being paid to oil discoveries: but the underlying story is that large finds are becoming extremely rare.[20] This is now causing attention to focus on natural gas discoveries, since oil decline will suggest to many that we should move our hydrocarbon dependency to gas. Indeed this shift to gas is already under way.

Since natural gas is now a critical resource for our world industrial system, in rather the same way as gasoline is critical to a car, we need to know how much we have got, and where to get more.

Hence, understanding where gas is should be part of general energy literacy, as should knowing why natural gas is sometimes found in different places from oil.

It may already be obvious, from the earlier description of the creation of natural gas, that the amount of "cooking" that the source rock receives will strongly affect how much gas there is compared with oil, as will the type of trapping system and the kind of organic matter. Higher heating levels generally mean more gas, but gas also requires better sealing or trapping; otherwise, it will leak away more easily and quickly. In the latter case, there may still be oil present in the rock. In a case where great heat has been present, along with good sealing, only gas may be found.

Once a deposit of gas is discovered, it is very important to estimate its size, especially since natural gas is getting harder to find, and more exploration is taking place in "frontier" locations such as the Arctic and very deep water,[21] where the extraction companies must try to assess whether its removal is economically feasible. Given that much of the gas is more than a mile beneath the earth, it is perhaps surprising that we discovered it at all.

The First Clues

In fact, it was those places that had "leaking" reserves that were the first clue to gas deposits in the earth. Some 2,500 years ago, the Chinese discovered such leaks and found that they were flammable and a good source of energy, in this case for getting salt from seawater by boiling it. They must also have discovered early on that not enough gas leaked out to be useful for serious energy purposes, so they bored shallow wells with simple percussion drill systems. They also faced the same transportation problem that we do now, and used bamboo pipes to carry the gas to their gas-fired evaporators. In some ways, little has changed in two millennia. The pipes are longer and made of metal, but natural gas is still mostly used for heating.

Other ancient peoples were, by Western technological standards, not nearly as enterprising as the Chinese. In many places, from the Caspian Sea through the Middle East to Southeast Asia, fiery gas leaks were looked on as miraculous and signs of God or gods, including the flame at the temple at Baku and possibly such incidents as the burning bush described by Moses in the Bible.[22] However, perhaps the most famous and sought-after "gaseous" temple in Western history, the oracle at Delphi, was not burning natural gas. It was certainly built over a small gas vent, but it was associated with some bitumen in the limestone below, producing "sweetly perfumed" and highly intoxicating ethylene, which was breathed directly by the swooning priestess before giving her notoriously ambiguous and obscure pronouncements.[23] Processed natural gas has no odor[24] and is not poisonous or intoxicating, though it does produce its own brand of economic mystification.

Extracting Gas

Although the Chinese used natural gas for working purposes, it was not until coal was used to produce methane in the late eighteenth century that sufficient quantities could be had at will to make it possible for the kinds of interior and street lighting constructed in Britain and the United States.[25] However, as with oil, the great early finds of gas were made in the United States. The first natural gas find was in 1821 in Fredonia, New York, with larger amounts coming in conjunction with the commercial discovery of oil some forty years later in Pennsylvania.[26] The giant finds of natural gas made in the twentieth century, often associated with oil, started with the great find at Spindletop, Texas, followed by many discoveries in Louisiana, Oklahoma, California, and then offshore in the Gulf of Mexico.

Natural gas, or at least what is called "conventional" natural gas, is extracted much the same way as oil, and often at the same time (since many reservoirs contain both forms of hydrocarbon). A vertical hole, or "well bore," is dug with an expensive drill bit made of very hard, often diamond-coated, metal. The drill bit is secured at the bottom end of an increasingly long string of pipe consisting of 30-foot[27] sections of drill pipe, called "stands," screwed together to form what is called the "drill string," which extends from the derrick[28] floor deep into the ground. After each 30-foot section of well bore is drilled deeper into the earth (called "downhole"), the main diesel motor that rotates the drill string and drill bit is disengaged, and another stand of 30-foot drill pipe is raised high in the air by the derrick and lined up over the exposed end of the drill string and then screwed on. The main motor is reengaged, and the entire drill string and bit begins turning again. It revolves at about sixty revolutions per minute, roughly the speed of an old gramophone record, and depending on the rock below, cuts about an inch a minute. In a deep well of 10,000 feet,[29] a drill string of 4-inch[30] drill pipe will weigh in excess of 155,000 pounds (or about 77 U.S. tons[31]). In the case of directional or horizontal drilling, the drill bit assembly at the end of the drill string has a precise bend, or "dog leg," that enables the drilling of a huge underground radius that eventually attains the desired horizontal angle and azimuth[32] of the well bore. Thus, the well bore on a horizontal well creates an L-shaped profile from the vertical top to the horizontal lateral that extends out and into the formation.

With land wells, these horizontal or lateral "kick-outs" are often 3,000 to 6,000 feet long and can vary in angle and direction throughout their length to strike the regions where oil and gas are expected, known as "target or pay zones." It is also common for a single well to have multiple lateral (horizontal) well bores, sometimes "stacked" one on top of each other at different depths or even "opposed," extending, say, east and west from the vertical well bore, if it is thought that oil and gas are present in different layers and in different directions. Once the drilling is finished, concrete is poured down to set in place a "casing" of large-diameter pipe to prevent collapse of the vertical well hole. A vertical well is then "perforated" with precisely placed explosive charges at depths where oil and gas are known to be. These explosives blow holes through the casing and through the surrounding cement, and about 12 to 18 inches[33] out into the formation.

The perforations allow the hydrocarbons to flow into the casing and up the production tubing to the wellhead, and then on to a "separator"

that breaks the entrained gas out of the oil. The gas then enters the natural gas pipeline system.[34] In the case of a horizontal well, casing is set
down the vertical well bore, but it is not necessary for the laterals or the
horizontal sections that intersect multiple fractures and fissures.
Horizontal sections enable "open hole"[35] production from all connected
reservoirs, and greatly increase production from certain kinds of reservoirs that have oil and gas in many vertically separated pockets.[36]

This is a description of a modern, highly accurate drilling operation, where the position of the drill bit is known to within two or three
feet,[37] even thousands of feet down and at high angles of drilling, and
where sophisticated blowout protectors at the wellhead greatly reduce
the chances of accidents and explosions. Life was very different for the
earliest drillers, however, especially those using percussion drills
adapted from salt and water drilling. For them, and even decades later
in the twentieth century, finding gas, which could explode, was often
a hazard, since the prize was oil, and geological understanding was
often rather primitive.

The extraction process described above refers to conventional gas,
which is that gas found in porous and permeable[38] structures similar to
oil, either on land or in relatively shallow water, and which flows naturally and is recoverable by traditional methods. Unconventional gas
will be dealt with later in this chapter, since it is invariably conventional gas—easier, cheaper, and available at much greater flow rates—
that is extracted first.

Separation Anxiety

When the price of natural gas at the wellhead gets very high, there is a
great temptation to leave the liquids in the gas stream, rather than
removing them and selling them separately. The reason is simply economic: when natural gas is worth more than butane or propane, why

pay the cost of removing them and make less money? The problem with this approach is that these liquids cause difficulties with the compressors and other parts of the distribution machinery and deprive industry of vital chemical feedstocks. If pursued long enough, this strategy will cause the supply to fall and the price of the liquids to rise, and there will eventually be more incentive to remove them. It is, however, a rather clumsy and unfortunate way to operate: it would be much better if the chemical industry knew in advance that liquids were going to be left in the pipeline and could either plan to pay more or re-schedule operations. It would be better still if the liquids were always separated as a matter of regulation. In a wider sense, the gas industry has found itself in many a production fiasco because it now relies almost entirely on price signals, which tend to be masked, time-lagged, or out of phase with the extraction process. "Big energy" is an industry that requires a lot of planning, just as it requires a lot of infrastructure.

The Final "Prize"

With the development of pipeline networks, the United States and then the rest of the world intensified the effort to find and extract more and more natural gas. Increasingly, natural gas looks set to become the great (and final) "prize" of the early part of the twenty-first century.

The twentieth century saw an extraordinary rise in the discovery and exploitation of natural gas, though much of the discovery was a by-product of the search for oil. Until recently, the United States held title to three records: it was the world's largest user of natural gas, the world's largest producer of gas, and, until the late 1960s, it also had the largest reserves. Since 2002, only the first is still true. The United States consumes far more natural gas than any other nation, but Russia now produces more gas and has more reserves than the United States. It is chiefly the U.S. fall in gas production that has prompted the need for this book.

The decline in gas supply, without any accompanying decline in demand, may turn out to be as disastrous for the world as the 1970 U.S. oil peak was. Neither of these events has appeared to be openly calamitous for the United States because of its immense military and commercial power and the power of the dollar as a foreign reserve currency and settlement currency for oil. But the world has suffered the curse of various U.S.-backed dictators on every continent, dictators whose regimes must guarantee a steady supply of oil to the United States.

All Western-style industrialized countries are already hopelessly addicted to natural gas, and, with the sole exceptions of Norway and Australia, either have declining or no gas production of their own. Rather than avoiding this foolish trend, all the rest of the world is following fast in growing gas demand. Thus, home user and policy maker alike should have an idea of whom they will be depending on for their gas in the future.

Reserves and the Three Ps—a Cautionary Tale

The next thing to say is that the following account of who has what natural gas reserves must be treated with considerable caution! Natural gas reserves are very difficult numbers to be sure of. There are several reasons for this, primarily financial, definitional, and political.

All reserve numbers are estimates of one sort or another. Before a drill bit finds gas flowing in commercial quantities it is never certain that the field will be economical. The Sable Island offshore deposits in eastern Canadian waters are an example of this. Even with commercial flows, and many wells, until a field is exhausted or abandoned, it is not possible to say with absolute surety what the ultimate recoverable reserve is. Therefore, reserve estimates must be made, using information obtained from drilling and either a lot of judgment (in deterministic estimates) or a lot of computing (for probabilistic estimates).

Many of the issues and difficulties with reporting the reserves of oil and gas are similar. The most important differences, however, are that the amount of gas that can be recovered from a deposit is usually much higher than for oil, as high as 80 percent for gas, but only 40 percent for oil. The great disadvantage for gas is that most of the world's remaining deposits are stranded, far from users, and often in very hostile environments. Since gas cannot just be loaded into the holds of a tanker, this imposes a much higher financial threshold to build the extremely expensive infrastructure needed to transport the gas. Alaska and northern Siberia, with reported sizeable gas deposits, are two clear cases of this problem.

There is no international agreement on what constitutes conventional or unconventional gas, or on exactly how reserves should be defined. There are three main terms for reserves: proved (or proven), probable, and possible— except in Russia.

Proved Reserves (1P)

Proved or proven reserves, often termed 1P in the industry, are defined by the Securities and Exchange Commission (SEC) as "those quantities of gas which, by analysis of geological and engineering data, can be estimated with reasonable certainty to be commercially recoverable, from a given date forward, from known reservoirs and under current economic conditions, operating methods and government regulations."[39] It is obvious that this definition depends on economics, technology, and politics, which means that proven reserves are not fixed. This is both fair and foul, because it allows for all kinds of nebulous and convenient interpretation for political and financial purposes. However, in recent years, the SEC has enforced its conservative accounting practices, so that in the United States oil and gas reserves may be slightly underestimated. Using methods known as probabilistic, 1P equates to P95, meaning that there is 95 percent certainty that the stated amount is both there and can be extracted.

There is much criticism of the SEC system, though some argue it is a good thing, since it encourages caution in reserve reporting. There are two main reasons for criticism of using only proved or 1P. First, it allows—indeed encourages—the practice of "reserve growth," in which petroleum companies report or "book" only what they are producing, which is another way of saying proved. This conveniently means that in bad years, when they don't discover much new oil and gas, they can draw on their existing reserves, which privately they know they have, but have not reported or booked. This makes it look as if they are continuously discovering new gas (or oil) and that a reservoir actually grows with time, often, it is suggested, because new technology makes ever more recovery possible.[40] Reserve growth is a practice that has allowed petroleum companies to hide the fact that in reality they are discovering less and less oil and gas. It is one of the reasons why so few people are aware that we are moving into the decade of global oil peak and North American gas peak.

The second reason for criticism of the "proved" only system is that most of the rest of the world follows a quite different practice, namely 2P, "proved and probable."

Proved and Probable Reserves (2P)

Fortunately, although the SEC does not count "probable" reserves, the USGS[41] has a term, 'inferred reserves," which is equivalent to "probable." An inferred reserve is one that is part of an identified economically producible reserve that "will be added to proved reserves in the future through extensions, revisions, and the discovery of new pay zones in already discovered fields."[42] When proved is taken with probable it amounts to the "probabilistic" equivalent P50, which means that there is a 50 percent chance that the amount will be recovered.[43] Reserve estimates published for most countries are 2P reserves (or discovery).

"Proved and probable" generally amounts to a reasonable guide as to how much gas will finally be extracted. If reported honestly, "proved and probable" is also largely immune to reserve growth, which means that 2P should only grow if real, new discoveries are made. If not, one should expect it to decline as it is produced.

Until recently, Canada used a reserve-reporting system similar to the one used in the United States. However, in 2003 it changed to a system more in line with world practice, namely 2P.[44] This should mean that companies should be reporting a one-time increase in their reserves, but that is not always happening. One reason for this is that there are exemptions for large companies,[45] who may be able to adopt the SEC rule of proved only (1P) if approved, and that Canadian reserves are mature, and therefore have little room left for "reserve growth" anyway. It is a little like locking the stable door after the horse has bolted.

All the Ps—Proved, Probable, and Possible Reserves (3P)

Finally, countries and corporations may report an additional third P, which becomes "proved, probable and possible," or 3P. Technically this can be described as "reserves where commercial productivity has not been demonstrated, and there is a 10 percent probability that at least the sum of the estimated proved plus probable plus possible reserves will be recovered."[46] In reality, this number is not helpful and may indeed further the process of avoiding the simple fact that gas deposits, though large, are limited.

Russia

For historical reasons, Russia has a completely different classification system. Since Russia is currently the world's largest producer of gas, a quick translation is required. The Russian system uses categories A through D to describe how ready a reserve is for drilling and development. Most reserve reports speak of ABC1, and this is generally regarded as being equivalent to proved and probable (2P), less about 25 to 30 percent.[47] Russia also reports C2, C3, D1, and D2, which may be equivalent to "possible." With a few exceptions, only the ABC1 numbers are important.[48]

The subject of reserve classification in the world of gas (and oil) is fraught with difficulties and probably will remain so. This is most regrettable and will become more so as the world moves into oil decline[49] and later gas decline, since when a vital resource gets low, it becomes very important to know exactly how little you've got left.[50]

Political Reserves

To compound this confusing situation with reserve reporting and definitions, gas, like oil, is now also a highly political substance, and countries and corporations rarely tell the whole truth about what they think is under the ground, even if they think they "know." The existence of what amounts effectively to two sets of books, publicly cited reserves (political) and technical reserves, is unfortunate.[51] Where possible in this book there will be an indication of the difference between what countries say they have and what skeptical analysts think they have.

The "technical" numbers are the ones that petroleum geologists report in considerable secrecy. They usually represent a much better estimate than the public numbers. The "public" numbers are the sizes of reserves that nations would like the world to believe they possess.

For U.S. corporations filing with the SEC, the public numbers are, ironically, as explained earlier, actually the most legal numbers, because the SEC only allows reserves to be booked if they have seen a production drill bit.[52] Most important, thanks to these reporting rules, as more oil and gas deposits are developed, the public numbers tend to grow over time. This produces a comforting and convenient illusion of ever-increasing reserves, often called "reserve growth," whereas the technical numbers tend to remain unchanged. Thus the technical reserves usually start higher than the public but are not subject to constant adjustments.

One of the reasons for this problematic practice of "reserve growth" is that countries wanting to increase their political bargaining power may have no commodity other than oil and gas reserves that can be used as collateral for loans. Corporations, on the other hand, often tend to downplay their reserves for tax reasons, in order not to prompt a country to nationalize the resource or, increasingly, to put other companies off the scent and avoid bidding wars for leases. Furthermore, for the corporation, this "dual number" situation has another useful spin-off, because it means that the public numbers can easily be manipulated to dupe economic analysts and Wall Street, who always want to see growth. These numbers allow for just the right amount of growth to suit whatever is in fashion with investors.

Unconventional Gas

Unconventional gas refers to natural gas extracted from coalbeds (coalbed methane or CBM) and from low-permeability[53] sandstone (tight sands) and shale formations (gas shales).[54]

The U.S. Department of Energy predicts that unconventional gas will become the most important source of domestically produced gas[55] when compared with conventional onshore and offshore production.[56] Whether the full expected rise in unconventional gas production takes place or not will become of vital concern to all North Americans (including Canadians and Mexicans), and to much of the rest of the world. In trying to gauge the future delivery path of U.S. unconventional gas production, it is noteworthy that there is an enormous upward revision in the U.S. Energy Information Administration (EIA) predictions from 1999 to 2003,[57] and equally interesting downward revision of conventional production. It is precisely the strong evidence that the conventional production decline is already turning out to be correct and that the badly needed compensating unconventional growth will fail to materialize that is underlying the U.S. gas crisis and the need for much more understanding and critical interpretation.

Coalbed Methane

U.S. production of onshore conventional gas really began declining in the 1980s, but it was covered up by the development of a host of large fields offshore in the Gulf of Mexico,[58] and by the increasing use of several kinds of unconventional gas, particularly coalbed methane (CBM), but also tight sands gas and gas shale. Much of the large increase needed in U.S. unconventional gas production has to come from CBM, which now produces about 10 percent of U.S. domestic gas, and accounts for more than 10,000 wells.[59] Reports from the Powder River, a major basin of CBM, that production output has ceased growing for the first time in eighteen years do not bode well for a great expansion of CBM. In any case, a huge increase in drilling is always required to increase CBM production because of the nature of the deposits.

Coalbed methane is found, as the name suggests, associated with underground coalbeds with the gas "adsorbed" onto the surface of the coal. It is thought that coalbed methane deposits may be able to store six or seven times the amount of gas for a given volume of rock compared with a conventional natural gas reservoir,[60] and it may be found at relatively shallow depths of 500 feet, though it can range down to 5,000 feet.[61] Shallow drilling is a benefit, but the drawback is that the reach of each well is not great, so that many more wells must be drilled for a given size of gas deposit than for a conventional gas reserve.

Coalbed methane gas is normally released from the coal seam by drilling a well and pumping off some of the water that fills the pores and cracks in the coal seams. An average well may produce as much as 20,000 gallons of wastewater a day for about 21,000 cubic feet of gas production.[62] Some wells only produce a quarter as much gas. Furthermore, a well may have to be "dewatered" for up to a year before significant gas production begins, though wells may begin to produce much earlier than that. Over time, as the coal seams are dewatered, more gas may be produced for less water pumped.[63] The amount of gas produced depends on a number of factors, including geology and whether there has been a lot of drilling for conventional gas and oil nearby, which is quite frequently the case. Conventional drilling and extraction in the proximity lowers the pressure in the CBM field and thus reduces output.

Methane is generally found wherever there is coal and in the past has been an unwanted and fatally explosive gas. It continues to kill thousands of miners a year, especially in China, which holds the unsavory reputation for the world's most unsafe mining practices. The United States, which has by far the world's largest coal deposits,[64] claims CBM reserves of only about 18 trillion cubic feet (Tcf), which is much less than Canada reports,[65] though Canada's coal reserves are tiny by comparison.[66] CBM production tends to be higher with the kind of big, fractured seams of coal found in the Powder River Basin in Wyoming, though the first significant coalbed-methane drilling occurred in the San Juan Basin in northwestern New Mexico in 1979. In the United States, Colorado, New Mexico, and Wyoming hold almost three-quarters of proven coalbed methane reserves,[67] but there are many other CBM-producing areas, including Appalachia and the Rocky Mountain region.

The United States has been producing CBM for two decades, but Canada only began to see major CBM efforts in 2003. Canadian CBM has been characterized so far by lower production volumes of gas, but also of less, or in some cases, no water production. Though still in its early stages, some areas of Canadian CBM production appear not very different from conventional gas production.[68]

The problems with coalbed-methane production in the United States are legion. First, many of the chief CBM deposits are federally owned but lie under private land. However, in some major CBM states, thanks to an arcane twist of law, the owner of the mineral rights trumps the surface owner, so that a developer can build roads, tear up land, start drilling, and site noisy compressors without even telling the landowner. Much of the surface land is owned by ranchers, and their land, livelihoods, and indeed lives are being ruined by the gas producers.

As the United States grows more desperate for gas, and as laws are passed to "streamline" the permitting procedure, this phenomenon is likely to get worse. In a sense, these unfortunate ranch owners, often Republican voters, are getting a taste of what the poor people of Texas and Louisiana who live near the giant refineries have been putting up with for decades, as they are overwhelmed time and again with toxic "incidents" and poisoned water.[69] Such residents have found themselves largely powerless.

The second and perhaps even more significant problem with CBM is the large quantity of wastewater produced, which was referred to earlier. This water contains many contaminants that increase its salinity and sodium[70] to abnormal levels.[71] When this water reaches ordinary soil, it will often kill the existing vegetation while encouraging noxious species. The discharged wastewater has a toxic effect on crop and range lands, especially when used as irrigation water. It has even been sprayed or "atomized" into the air in the hope that it will evaporate, but during the winter, this mist can freeze into extraordinary mini ice mountains,[72] which resemble structures from Monument Valley.

Despite the banning of "direct stream discharge" for new wells, many CBM producers are still allowed to discharge straight into stream channels under "grandfathering" schemes, which allow old wells to continue polluting. Wells in the Powder River Basin in Wyoming can produce up to 28,000 gallons[73] of contaminated water a day, and there may be as many as three wells for 80 acres,[74] which can mean much more than 50,000 gallons[75] a day. If this water isn't dumped into rivers or reinjected underground, it is collected in enormous wastewater "impoundments," which cover the landscape in some major CBM regions. The industry claims that some of the water is drinkable, and there are efforts to reduce its effects by cleaning it. Given the less than sterling environmental record of the oil and gas industries in the United States and around the world, it must be assumed that they will do the least possible, and only then when there is media attention or extremely well-organized and determined citizen opposition.

Tight Sands and Gas Shale

The other unconventional forms of gas currently in production are tight sands gas and gas shale. Tight sands gas, which means that the gas is held in reservoirs where flow is much slower, is considerably more prolific than coalbed methane but requires much more drilling. In the United States, production appears, at least for the moment, to have flattened off. In terms of amount produced, gas shale represents a much smaller proportion than either tight sands gas or coalbed methane and has not grown nearly as fast as CBM. The EIA predicts that tight sands gas will play a much bigger role in the future, which is not good news for the residents or the ecosystems of the Rockies where more than half of the tight sands gas is located.[76] The increased drilling needs of tight gas and its locations in sensitive areas are major reasons why the U.S. Congress came under increasingly strong White House and industry pressure to make access and permitting easier. This pressure was crystallized in the September 2003 advisory report from the National Petroleum Council (NPC), which advised that more supply could only be had by opening as much new potentially gas-rich land as possible. The unpalatable truth is that the U.S. conventional gas supply peaked back in 1973, and only the rise of offshore production and unconventional gas has made up the difference.

Methane Hydrate

There are many who believe that we needn't worry about gas supplies because there are vast quantities of methane hydrates in the oceans and at the poles. When prices rise high enough, the argument goes, the technology for commercially and perhaps even safely recovering this resource will be found. There is disagreement about almost all aspects of methane hydrates. Some, including the U.S. Department of Energy, claim that there is a great deal of methane hydrate, possibly many times the amount of fossil gas, while others, with very good evidence, suggest that these estimates are exaggerated, and that there are many important unanswered questions and contradictions in the case being made for methane hydrate as a large-scale substitute for conventional natural gas.[77] It is also noteworthy that the petroleum industry has shown no enthusiasm for methane hydrate as a primary source of energy.

There are a number of serious problems with methane hydrate. For the producers, no feasible way has yet been developed of recovering the gas hydrates despite great efforts by various governments, and at the time of this writing (February 2004), methane hydrates provide precisely none of the world's gas supply. There has already been a great deal of activity to find and develop methane hydrates, especially following the various oil and gas shocks of the 1970s. These efforts did not produce any commercial results. Once again, there are renewed efforts to research and develop methane hydrates, such as looking at methane vents in the Gulf of Mexico,[78] Hydrate Ridge off the Oregon coast, and the Nankai Trough, near Japan.[79] Whether newer ventures will fare any better or worse than previous attempts, it is too early to say.

Petroleum companies have taken interest in methane hydrates in one way. There is concern that seabed drilling can disturb the hydrate structures and lead to seafloor subsidence, especially slumps along continental margins, which could cause serious safety risks to the drilling crew and problems for the drilling process itself.[80]

A problem on a much larger scale that many scientists worry about is that mistakes with methane hydrates could trigger colossal releases of methane. Such releases in the past appear to correlate with large and destructive global warming events. It is not possible to state that no one will ever solve the production problems and commercially capture the gas in methane hydrate, but large extra methane releases would only worsen the global warming problem. While many petroleum geologists dispute the carbon dioxide connection to global warming, methane is more than twenty times as harmful a greenhouse gas. Given the sad record of disregard for human and ecological safety that has characterized the petroleum industry for more than a century, it would be foolhardy to have any confidence in the industry managing to avoid large-scale methane leaks. If methane hydrates were to be produced commercially, given that numerous scientists offer evidence of a strong connection between fossil-produced carbon dioxide and global warming, the prospect of being able to burn almost unlimited quantities of natural gas should be enough to galvanize the world community into demanding that we not go down that path, even if we can.

Finally, providing a new kind of energy, which methane hydrates would be, at vast scale takes decades to implement. In this regard, it is supposed by some that a kind of Apollo Moon-shot program might be the answer regarding the recovery of methane hydrates or possibly some other energy form. There are several powerful objections to this idea. The Apollo program was a political program and government funded at a time when the United States had a lot of spare cash. With the sole exception of the military budget, the United States will not spend government money on any project that could theoretically be carried out by the private sector—even when corporations quite obviously won't because the project is not profitable in the short term, or too uncertain in the long term. Neoliberal regimes, such as those that have been in power in all English-speaking countries for two decades or more, just won't invest in projects for the public good.[81] It is against their philosophy. There are some who think the United Nations should undertake such a huge project. However, it is clear, by mandate and practice, that the United Nations is not suitable for this task.

The second argument is more structural, though equally damning. The Apollo program, while clearly a huge undertaking, was aimed at doing one single thing: getting a person on the moon. This kind of task is similar to making a film, where everything has to work perfectly, but only once, and money is no object. The project of substituting the planetary energy grid involves an utterly different kind of philosophy, as well as technological and systemic challenges untold orders of magnitude greater. It doesn't mean that we couldn't achieve something impressive, given the will, the time, and the money, but currently, we have none of those things.

To conclude this argument, energy is what allows us to destroy the planet, so perhaps we should actually hope we don't find free energy systems or learn how to unlock the methane hydrates.[82] We have demonstrated beyond a doubt that the more cheap energy we have, the more damage we do. We have shown that guided by economics, we are utterly incapable of controlling ourselves, most certainly at the level of global corporations or government, especially in the most powerful countries, where the two are now virtually the same thing. It might be

worth remembering that Mussolini referred to the seamless conjunction of government and corporation as Fascism.[83]

For all the reasons above, methane hydrates should be barred from
exploration and attempts at exploration, much in the way that we have
mandated the Antarctic off-limits to hydrocarbon exploration. (The
fact that there is probably nothing to find there undoubtedly makes
that injunction easier to enforce.)

Man-Made Gas

Coal has been used in the past to make gas. In the United States, this
gas was similar to natural gas, though dirtier and more poisonous.

In Britain, the gas produced was called town gas and typically con-
sisted of hydrogen, carbon monoxide, and methane. It burns with less
energy than natural gas, and the carbon monoxide component is
intensely poisonous, causing asphyxiation if it escapes. Carbon
monoxide is odorless and colorless, and since it is about the same density as air, it will not rise but diffuse slowly into the surrounding air,
which adds to its danger. Given the countless accidental deaths from
the use of town gas, the relative nontoxicity of natural gas is one of its
many advantages.

Methane Creation

Methane, the main constituent of natural gas, is being produced all the
time. Human beings produce it and so do most animals, especially
cows and other ruminants, as well as rice paddies. So much biogas is
produced from cows and rice fields that both can be considered a factor
in the creation of greenhouse gases. Of course, it is rather hard to capture the gas from moving cows or vast rice paddies. Methane is also

produced in landfills, where it is a nuisance until captured. Methane is also generated in compost heaps. There are serious efforts underway to harness the gases from the dung of cows and other animals, which, when mixed with organic "waste" matter, can be used to generate "biogas" in a methane digester.[84]

A methane digester is a device that converts animal feces, along with other organic waste, into biogas and leaves behind a sludge residue. The digester does this by collecting the waste, keeping out the air, and warming the material to a constant temperature of up to about 100°F.[85] This encourages microbes to break down the organic material, producing biogas[86] as a by-product, though as far as the microbes are concerned, the biogas is a waste product. Biogas contains anywhere from 50 to 80 percent methane, 20 to 50 percent carbon dioxide, and small traces of hydrogen, carbon monoxide, nitrogen, oxygen, and hydrogen sulfide. It always contains a lot of water vapor. Biogas burns to give about half as much energy as fossil methane.

Reports of biogas output and efficiency vary, and scale of operation can make a difference. In the United States, with electricity prices at less than ten cents per kilowatt-hour, biogas production for generating cost-effective electricity requires manure from more than 150 large animals.[87] A farm of 400 milk cows may be able to provide about 100 kilowatts of electricity, enough to power 75 average U.S. homes.[88]

As with any energy source, there are benefits and drawbacks. Digesters could certainly help produce a lot more methane than they do currently. In 2002, there were only about 40 digesters in use on American farms,[89] and more than 400 in Europe. The gas itself tends to be corrosive unless specially treated, which means that it is better to use it directly at the point of production. If converted into electricity, then most generator systems use the power grid to lock the frequency and voltage correctly.

In the United States, one of the unfortunate factors about methane digesters is that they work much better in conjunction with industrial factory farming. The system of factory farming should never have been introduced in the first place, for several powerful reasons, including its increasingly hideous treatment of animals, its fostering of more diseases and consequent overuse of antibiotics, and its high fossil-fuel requirements. However, methane digesters do at least help return more of the animal wastes to the soil and generate useful methane in the process. Ironically, the antibiotics in the animal feed may kill the microbes vital for the digestion process.[90] In Europe, however, there are active efforts to use less malignant inputs such as household organic waste, grass, waste vegetable oil, and many other non-factory-farm products. Both from an ethical and energy infrastructure-investment stance, this is a much more sensible strategy.

The problem with any sort of new gas production is the quantities we demand. Even if we managed to contain every molecule of new biogenic methane, we would have only a fraction of what we now use in terms of fossil natural gas.

Landfill Gas

Another way to produce methane is to collect landfill gas. Produced by the decomposition of organic waste, landfill gas is, however, much more problematic than biogas. It invariably contains materials that when burned produce a range of very serious toxins, such as dioxins and furans. In the United States, landfill gas is being promoted as green and renewable. There is certainly no shortage of landfill waste, but problems of the toxins appear to be very difficult to solve, since dioxins and furans are formed at a very wide range of temperatures, both inside and just outside any containing smokestack.[91] It would help greatly if the organic material were separated first and digested without the presence of the halides (fluorides, chlorides, bromides principally) and other organic chemicals, like benzene, which appear in our industrial waste, and are the culprits in the formation of the toxins. Unfortunately, to date the U.S. Environmental Protection Agency (EPA) has not been willing to consider separation. In fact, it would make far more sense to separate all the organic, mainly food, waste at the consumer end and digest it in local systems. As much as 60 percent of household waste is food. The other clear, and apparently unpalatable, policy implication is that we should produce far less waste in the first place, most especially the kinds of waste that are either toxic to start with or that produce toxins when incinerated. Existing efforts to reduce the waste stream at the local level are admirable but represent a tiny fraction of what needs to be done by industrial society in general.

Despite the above possibilities for making new gas, with current mammoth levels of consumption, fossil gas is our only option for carrying on with business as usual.

3

Moving Gas

Elusive Energy

Even though natural gas is made of the same basic carbon and hydrogen that oil is, the fact that it is a gas makes it fundamentally different when it comes to handling and transporting. In fact, like any other gas, you can't handle it; that is the point. Gases occupy anywhere between five hundred and one thousand times as much space as their equivalent liquid form. This means that moving large quantities of gas must be done either by continuous means, such as a pipeline, or discretely in single containers, in which case something must be done to shrink the volume dramatically. The only two means we have of shrinking are compression at normal temperature or cooling at normal pressure until the gas turns to liquid, a process known as liquefaction. Compressing a gas always requires a lot of energy, and then a special storage vessel that can withstand the pressure without rupturing. In the case of a flammable substance like natural gas, there is the added risk of an explosion. Liquefaction of some gases is easy. Propane and butane become liquid relatively close to normal temperatures and pressures, which is why they are so useful in so many applications, from camping stoves to ordinary cars and trucks. Natural gas is another matter. It must be cooled to about $-260°F$ ($-160°C$). That requires a lot of effort and a lot of energy. Then it has to be kept cool, which requires specialized insulated containment vessels.

All of this is easily within the bounds of modern technology, but it is very expensive. In general, the continuous, pipeline method of gas transport is clearly preferable, because it is cheaper and, well, continuous. Electricity has something in common with gas in this respect: we need it in a continuous stream, and on demand. This dependence on fixed, earthbound pipelines means, however, that natural gas, certainly in North America, has required what is called a "regional" market.[1] There are undersea natural gas pipelines such as between Norway and Britain, but they are not very long, and they are even more expensive than land-based pipelines. If they rupture, repairing them is not a pleasant task, or a cheap one. The only reason to use anything other than pipelines to transport gas is the presence of a large body of water in between the producer and the user. This is the case already for Japan and Korea, who receive gas from Indonesia, and is increasingly so for much of the rest of the world, including the United States. In the first decade of the twenty-first century, transporting natural gas as a liquid is set to grow very quickly, because the largest users (the United States, Canada, and the United Kingdom) have declining supplies, but globally others still have gas to spare, if only they can get it to those that want it. Since the gas must still reach the ocean-going LNG tanker before it can be shipped, gas pipelines will also see intense growth. The land connections between Russia and Europe and China also ensure a big future for pipelines.

Pipelines

It is important to understand the role of pipelines, since these are what made the use of natural gas so widespread in the twentieth century, particularly in the United States. In several stages, beginning in the early 1800s and expanding vigorously in the twentieth century, a natural gas pipeline network comprising millions of miles of pipes (if one includes the smaller diameter, local distribution system) was developed and constructed across the United States. In some ways this resembles the federal highway system. As early as 1785 in Britain and 1816 in Baltimore, Maryland, pipelines were laid to convey gas made from coal to streetlights and houses. It was soon realized that natural gas straight from the ground was safer and cheaper than gas made from coal, and as mentioned in chapter 2, the first natural gas well was drilled in 1821. In 1859, like so many after it, the first commercial oil well in the United States also found natural gas. This gas, again mainly for lighting purposes, was brought to the nearby town of Titusville, Pennsylvania, by a 5.5-mile pipeline.

By the time of the first "long" pipeline in 1891, linking Chicago with wells in central Indiana, 120 miles away, electricity was already displacing gas as a lighting agent. Natural gas producers at that time had to find new uses for their product. More efforts at a pipeline network were made in the 1920s, but it was after the Second World War that enough technical progress had been made to realize the great pipeline system that now covers much of North America. These advances included improvements in welding, pipe rolling, and metallurgical techniques that produce much more reliable pipes. The boom in pipeline construction lasted until the late 1960s.[2] Now there are more than 280,000 miles of large, long distance "transmission" pipelines in the United States. These pipes vary in size from 20 inches to 3.5 feet in diameter. They require compressors every 70 miles or so to maintain enough pressure to keep the gas moving.[3]

The U.S. gas pipeline system encouraged the use of natural gas to grow to over 23 Tcf in 2002. It is this meteoric growth and resultant continuous demand for gas that is posing severe new problems not just for the United States, but for Canada too, since it is so closely—and it will find regrettably—connected to the U.S. gas market. The situation

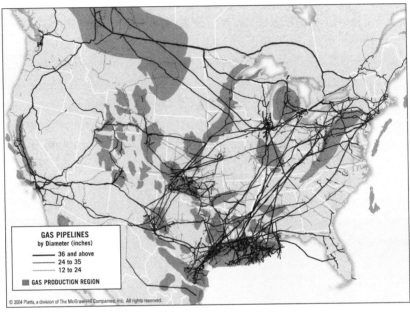

Figure 3.1. The North American natural gas pipeline system. *Source: Platts and POWERmap.*

developing in the early part of the first decade of the twenty-first century has serious social, economic, and security implications for North
America, but it should also serve as a warning to the rest of the world.

Pipelines are important both in facilitating the transport of gas to
the user and in holding back production. What this means in practical
terms is that if a large natural gas field that may be able to produce, say,
100 million cubic feet of gas a day has a total pipeline capacity connected to it that is only capable of carrying 50 million cubic feet a day,
then production will be capped at that lower level. In fact, this has a
considerable advantage, since it extends the life of the field, and can
produce a much flatter production curve than an oil field, so that gas
flow may resemble a reasonably reliable river. The problem with this
comes at the end. If the size of a field cannot be accurately determined,
then it will be hard to say when production will decline. When production does begin to slow in a gas field, it tends to do so very rapidly

and without the timely warning signals that oil fields typically offer. This gives rise to the term "gas cliff," with the analogy to a very steep and rather unfortunate drop.

In the United States, which has so many small gas producers, pipelines matter a lot in the natural gas system because they are the real source of control, not the wellhead.

Pipeline Problems

For the natural gas user, pipelines have many benefits. In the main parts of the industrialized world, oil and gas pipelines have a fairly good, though not unblemished, safety record. In the producing regions, whether in North America or in poorer nations, the same cannot be said. Problems range from cases of poor construction by petroleum companies in Alaska to nonaccidental deaths in Colombia, which are accepted by the petroleum industry as a cost of doing business. The rush to put pipelines into Africa has been attended by military violence, with accusations of operations being covertly paid for by production companies. There have also been disastrous pipeline accidents, either through poor construction or because impoverished local people, through whose land a pipeline may run and who get no financial benefit from it, rupture the pipelines and try to collect the leaking fuel either for themselves or for thriving black markets. One of many such incidents in the oil-rich Niger Delta killed over five hundred people in 1998.[4] This is not to say that all petroleum companies act with callous disregard for human life, or that most of those who work for them condone this behavior.

When oil pipelines are ruptured, either by accident, poor engineering, or by explosion, it is obvious that the likelihood of pollution into the soil and rivers is very great. In contrast, natural gas pipelines are, by the very nature of methane gas, not going to produce such obvious ground-based pollution. However, for as long as the pump pressure is not turned off, gas will escape from the rupture. Raw methane gas is, as mentioned earlier, a very powerful greenhouse gas, and it is also explosive. Sometimes oil and gas pipelines run side by side, as in Iraq. When an oil pipeline is attacked, the gas line may be hit as well or by mistake. In the new world order of energy desperation now unfolding, there is no reason to believe that pipeline attacks will not continue and increase. Ironically, despite being made of the highest quality steel, they are the ultimate "soft" target.

Pipelines are particularly easy game for those with grievances against the producing corporations or the government in question. Aggrieved peoples seem to come from almost every country outside the industrialized world. Attacking pipelines has, for instance, been a constant, and from the rebels' point of view, rather successful tactic in Colombia for many years. With the 2003 U.S. invasion of Iraq, it has become popular (and effective) there too.[5]

Another pipeline issue is the way native peoples and tribes are dealt with by the corporate producers and the governments that allow the extraction. When a native tribe or a village is in the way of a pipeline, they will often be forced to leave, sometimes at gunpoint, with little or no compensation. In Burma there are cases of villagers being forced into slave labor.[6] How a community can be "compensated" for the loss of its ancestral home is highly questionable in the first place. In the worst cases, people have been murdered. The oil companies, of course, would never commit murders directly, not least because this might detract from their newfound green marketing tool, "corporate social responsibility." Instead, the local or state militia carry out the work. The examples are legion and well documented by many human rights groups.[7] A representative selection of nations might include, for instance, Irian Jaya and Aceh in Southeast Asia, Colombia and Ecuador in South America,[8] Chad and Cameroon in West Africa. But these places are far enough away from white and Western eyes so that corporations can quite safely deny everything, knowing that the indifferent consumer, most interested in getting the cheapest price, generally doesn't want to know anyway.

Engineering Difficulties

As mentioned in chapter 2, historical sources suggest that the earliest gas "pipelines," 2,500 years ago in China, were made of bamboo. In Britain and the United States, in the early nineteenth century, wooden and iron pipelines were constructed. They were likely less than ideal, especially the wooden ones. As mentioned, it was not until some decades into the twentieth century that sufficient progress with pipeline techniques had been made to allow the widespread construction of a reasonably efficient and safe system of distribution. Yet, it appears that pipelines are still a weak link, both in North America and all over the world. No doubt they could be made more reliable, but at much greater cost. Pipelines are expensive to build, often wildly so in harsh, so-called "frontier" regions, such as Alaska and Siberia. Problems with rusted and leaking oil pipelines are common worldwide. Ruptures and leaks in gas pipelines are more difficult to detect than in oil pipelines, since one cannot see any green-brown or black liquid collecting on the ground. With the increasing use of gas, unless there are extraordinary and expensive improvements in pipeline quality, safety, and security, there will be more and more leaks of natural gas.

Deregulation

As more natural gas pipelines were built in the United States in the 1920s, and the U.S. economy entered one of its great financial bubbles, gas demand increased rapidly. The prices and supply were mainly controlled by the pipeline companies, and that led to price instabilities and worries over monopoly practices. The first regulation of natural gas in the United States occurred in 1938, when the Natural Gas Act was passed.[9] It created the Federal Power Commission (FPC) and charged it with stabilizing gas prices and supply. The FPC also set standards for the safety of compressors and pipelines. The 1938 act was very effective at helping create a growing market for natural gas by establishing federal control of interstate pipelines. The act did not regulate wellhead prices in many cases until the Phillips case of 1954. The ensuing ruling ensured that prices were set to allow producers to cover their costs and add a profit on top.[10]

For several decades, the regulation system ensured stable prices and supply for both user and producer and helped create the environment in which the pipeline network grew after 1945. However, the rather rigid design of the regulation system meant that it could not cope with large external changes, such as a huge increase in demand. With wellhead prices kept low, there was decreasing incentive on the part of producers to explore and produce new fields.[11]

By the same token, users turned ever more to this cheap and reliable source of energy and chemical feedstocks. It was the 1973 oil shock, however, that began to expose the systemic flaws. As oil prices rose, gas became even more attractive, and consumption ballooned, while prices stayed flat. Inevitably supply could not keep up with demand: U.S. conventional gas production reached an all time peak in 1973, which it has never seen again. With the cold winter of 1976–77, a full-scale natural gas crisis arrived.[12] Since this was the beginning of a new era of economic fundamentalism, rethinking or readjusting regulation was considered ideologically unacceptable. Instead, total deregulation was the path chosen, as shortages and price instabilities "proved" that regulation was not best for consumers.[13] Ironically, the 1978 Natural Gas Policy Act (NGPA), though designed to lead to full deregulation, caused considerable instability itself. Undoubtedly, the result of the act was more gas supply: user prices went up considerably, and that helped to quell demand. However, many producers were not ready for the full force of the free market, and difficulties, especially with interstate pipelines, remained. The Federal Energy Regulatory Commission (FERC) in 1985[14] decoupled natural gas supply and transportation, so that one no longer had to own the pipeline to ship gas. The Natural Gas Wellhead Decontrol Act of 1989 eliminated price controls on the wellhead sales altogether. Finally, in 1992 FERC let the pipeline companies create a market determined by how their capacity would be used.[15]

The years following the NGPA of 1978 saw a great increase in drilling and production activity, reaching an extraordinary peak in 1981, when over 4,500 rigs or separate drilling operations were sinking new wells either for exploration or extraction. For many years, this largely free-market system kept pace with demand. However, there are at least two important things to understand about what happened and what began to go seriously wrong in the new millennium.

The problems of regulating or deregulating the natural gas industry probably have no ideal solution. In the United States, there are thousands of small producers, and once gas supplies spread from being local or municipal to being intrastate, and then interstate, the problems of governance became immense. This situation has its parallels with the closely related business of electric-power generation, which has begun to break down seriously across the industrial world, especially in 2003. California led the way with its catastrophic power crisis in 2001. Although this appeared to be an electrical problem, there was an underlying component of natural-gas supply difficulties. It became apparent that deregulation in general had allowed powerful, mendacious, and highly unscrupulous corporations such as the now infamous Enron to charge extraordinary energy prices to captive or unknowing customers. This situation was exactly what the first Natural Gas Act of 1938 had been designed to prevent. To help understand both the interconnections between gas and electricity, and the gas situation more generally, there will be more about the California crisis in chapter 4.

What happened in California and in North America after 2000 should be of the greatest interest both to North Americans and to the rest of the world, since the world once again, as with oil in 1970, has an example to avoid. In North America, now that the production of conventional offshore and onshore gas is in decline[16] even more pressure is being applied for deregulation so that new lands may be opened up for drilling and extending the reach of unconventional gas production, such as coalbed methane. Whether this will solve the North American gas crisis will be examined in chapter 5.

Canadian Tar Sands

Does anything stand between the wellhead and the natural gas user? Conventional economic theory states that substitution is always possible. When one thing runs out or gets expensive, there will be something else that can be used instead, ad infinitum, world without end. So it is that Canada has been a fine and apparently limitless source of substitution for the United States for almost everything that it has used up with reckless haste. When the rate of U.S. oil extraction peaked in 1970,[17] and especially after the oil shocks of the 1970s, Canada became one of the key exporters to help the United States reduce its dependence on Middle East oil. Depending on local conditions, by the year 2000 Canada was often the number one exporter of oil to the United States.

Unfortunately, Canada's conventional oil reserves, while relatively large for the number of people in Canada,[18] were small by comparison to America's thirst. Too small. Conventional oil production, most of it coming from Alberta, began shrinking with the arrival of the twenty-first century.[19]

Full attention then turned to the Canadian tar sands, also called oil sands by those in the Canadian energy extraction business. Unfortunately, despite textbooks to the contrary, public relations language doesn't describe reality and certainly doesn't change physics, chemistry, or geology. The tar sands are not conventional liquid oil; they are literally a mixture of tar or bitumen and sand (and clay). It is a very difficult and expensive, highly polluting, and energy- and water-consuming process to extract the bitumen from the tar sands and turn it into synthetic crude oil. Tar sands only become interesting when reserves of easier forms of oil start to become scarce and expensive. This is clearly beginning to happen in the new century, as evidenced by the immense effort being expended on the Canadian tar sands.

Tar sands are formed when oil deposits lose their light components (such as conventional oil) through the actions of bacteria, water-washing, and evaporation.[20] There are several major deposits of tar sands in the world, but the Canadian deposit in the Athabasca region near Fort McMurray is by far the largest. It is hard to be remotely accurate in this matter, but there may be the equivalent[21] of somewhere between one and two trillion barrels of oil in this frigid and unforgiving place. That may well be more than all the oil that remains to be commercially extracted on the planet. But it isn't liquid oil. In the biting cold of northern Alberta, it's actually solid bitumen, and that makes all the difference. White Westerners have been aware of the Athabasca tar sands since at least 1875 and have been trying to extract it commercially ever since.

At the end of 2002, thanks partly to technological developments, but mainly due to backstage lobbying, a considerable part of the Canadian tar sands deposit was reclassified as conventional oil. This has had the dramatic effect of raising Canada overnight from being number twenty[22] to number two in oil reserves, behind only Saudi Arabia. However, while it might appear to reduce the significance and power of the Middle East oil producers, it does nothing to change the fact that current tar sands production processes consume an enormous amount of natural gas.[23] So large in fact, that it is suggested that the

Mackenzie Delta gas fields, said to have at least 9 Tcf of gas, which are not likely to be in production before 2010, would be consumed by the tar sands before they ever reached the United States. This is extremely important, since many U.S. energy specialists believe that this Arctic gas will help reduce U.S. supply problems. The Canadian National Energy Board predicts that by 2010 tar sands production will be well over one million barrels a day,[24] requiring at least 1 billion cubic feet (Bcf)[25] of gas per day to produce. Some in the industry have suggested that if all the tar sands projects were in operation that would consume 2 Bcf per day, which amounts to more than a quarter of current Canadian consumption.[26]

The situation in Canada highlights how complex the North American natural gas situation is. There are many problems large and small, often appearing as "mere details," yet any of which may turn out to be of great significance. In 2003, Albertan gas producers were ordered to stop extracting from two percent of their wells in the Athabasca region because of fears that their operations were lowering the pressure on the deeper bitumen deposits and making it harder to extract them. It was determined that the bitumen was worth more than the gas, but many gas producers did not see it that way.

There have been many other problems with tar sands development,[27] nevertheless, according to the petroleum industry a great deal hangs on the Canadian tar sands, and therefore natural gas must be made available for its extraction and production.

On the other hand, this same gas will be desperately needed both by Canadians and Americans as gas production declines across the continent. This is one of the very large issues that must be understood when weighing policy and personal decisions regarding converting to or from natural gas. Evidence and history suggest that moving away from natural gas dependence to renewables and using less is the only sensible short- or long-term strategy, for all North Americans.

Liquefied Natural Gas

Liquefied natural gas sounds simple in theory, but in practice it is not. If natural gas is cooled repeatedly, until it reaches about −260°F (−160°C), it will become a liquid at normal atmospheric pressure. The reason for doing this is that the liquid is about six hundred times denser than the gas, which means a lot more can be carried in a given volume, such as a ship. The trouble is it must be maintained at that temperature, which requires special vessels and plenty of energy. It also requires a great deal of energy to get it to that low temperature in the first place. The principle of LNG was demonstrated almost a hundred years ago,[28] but the industry didn't emerge at scale until the 1970s, when Japan began trying to reduce its reliance on Middle Eastern energy by moving to natural gas from the Far East.

At first glance, shipping natural gas seems a good idea. Long pipelines have practical and economic problems over 2,000 miles or 3,000 kilometers, which means that as much as 60 percent of the world's largest gas reserves are inaccessible for human use via pipeline alone.

LNG Train and Energy Costs

LNG requires a complex and extremely expensive infrastructure, called an LNG train, to connect production fields to the consumer. This train consists of a liquefaction terminal and pipelines in the producing country, colossal double-hulled cryogenic tankers, and regasification terminals in the receiving country. The whole train costs anywhere between $3 billion and $10 billion.

There are a number of problems associated with the difficulty of liquefying LNG and containing the cold liquid. Despite many advances in the science and technology of cryogenics (literally "making frost"), there are some limitations imposed by physics. Making any-thing very cold on our relatively warm planet takes a lot of energy. This may seem paradoxical, since one is taking the energy out of a sub-stance, but it is the heat or coldness relative to the surround that mat-ters. If one starts with a standard 1,000 cubic feet of gas, the process to make it into a liquid takes anywhere between 80 and 130 cubic feet, depending on the type and age of technology used, the outside temper-ature, and quality of feedstock.[29] The liquid is then pumped onto a ship, where it must be kept cold. To keep the methane in liquid form, some of it is allowed to boil off (though it is not lost), and this gas is used to power the ship itself.[30] The amount of "boil-off" per day is also rather variable, but it is within the range of 1.5 to 2.5 cubic feet per day (per 1,000 cubic feet of gas). Obviously the final amount of gas that boils off depends on the number of days at sea, and thus the dis-tance traveled. At 19 knots,[31] an LNG ship can cover about 500 miles a day, and for each 1,000 miles of water covered, about 4 cubic feet of gas boil off and are burned by the engines.[32] Finally, at the receiving port, more energy is used in the process of turning the liquid back into a gas, called regasifying. This takes about 10 to 15 cubic feet of gas. If we apply this to a shipment from Qatar to the eastern coast of the United States, a journey of about 7,000 miles, transporting each 1,000 cubic feet of gas will consume about 15 percent of the gas, or 150 cubic feet per 1,000 cubic feet of gas transported.[33]

Costs and Liquidity

Regarding the costs of new LNG infrastructure, even the very pro-business energy analyst Daniel Yergin has questioned the ability of the United States to find the $200 billion that may be required.

This "liquidity problem" will likely be further compounded by wider global needs during the next three decades to enhance and maintain world energy infrastructure, estimates of whose costs range as high as $16 trillion.[34] Again, even pro-market analysts have suggested that this will further strain the world's capital markets, and will not do anything to improve America's dangerous long-term multitrillion dollar debt.

LNG Accidents: The Price of Doing Business

The amount of gas consumed in LNG transport is an unavoidable part of the cost of doing business and is essentially an economic problem. A more serious problem is the danger of an LNG tanker exploding. The most well-known and tragic example of an LNG accident is the Cleveland disaster of 1944, when a new steel tank failed, spilling liquid natural gas onto the streets and into a storm sewer system. The resulting fire killed 128 people.[35] Lessons were learned from this incident, but on Staten Island in 1973, an explosion at an LNG facility killed 37 men working there.[36] Once again, the industry said that lessons were learned.

Then on January 19, 2004, in Skikda, Algeria, a powerful explosion destroyed three of the six LNG trains and badly damaged a nearby LNG loading berth, leaving at least 24 dead, 74 injured, and 7 people missing. Algeria is the world's second largest exporter of LNG, and the whole Skikda terminal supplies about a quarter of that country's LNG export. The Skikda LNG terminal was revamped and comprehensively checked by Halliburton in 1999. The work took two and a half years, and Halliburton said in October 1999, that the "Revamp Project has passed all performance tests."[37]

Within hours of the Skikda explosion, the LNG industry rushed to say that it wouldn't make any difference to U.S. LNG plans, and that the explosion had been caused by a steam boiler not directly linked with the LNG itself.[38] Even so, an industry journal, *Energy Intelligence*, wondered whether this would be the Three Mile Island of LNG. Given the coming shortage of natural gas in the United States, and the increasing imported gas needs of almost every other nation in the world, it seems highly unlikely that this, or perhaps any other cataclysmic event, unless possibly it were of Chernobyl proportions, will prevent the development of a large number of new LNG terminals, both in the United States and elsewhere. However, U.S. citizen-activists immediately seized on the Skikda explosion to bolster opposition to LNG terminals, at least onshore near population centers.[39] Some months after the explosion, a new report said that the cause was directly related to a gas leak. This has further hardened citizen resistance to LNG terminals across North America.

Despite the more serious cause, the Skikda explosion was still judged to be an accident. However, not two months before this, in early December, 2003, guerillas from the Islamic Salute Army (AIS) sabotaged a compression station on the Hassi R'Mel-Arzew gas pipeline, destroying two out of the three lines.[40] Hassi R'Mel is Algeria's largest gas deposit and supplies one-quarter of its gas production.[41] The LNG exports to the United States in the 1970s were also from Algeria, though at present the United States receives only a small quantity from Algeria.

The recent events in Algeria should in their own way be taken as seriously as Three Mile Island. Until Skikda, the LNG industry was quick to claim a very good safety record, notwithstanding the disasters in Cleveland and Staten Island. Skikda shows that it is literally a fatal mistake to say that no accident, be it sabotage or human error or unlikely engineering failure, will ever happen again, especially now that the United States, so dependent on foreign energy, has created so many enemies for itself, partly because of its pursuit of that very energy. LNG terminals and tankers contain millions of gallons of highly flammable methane. There is still much debate in the industry about what would happen if a huge modern tanker or LNG import facility failed in some catastrophic way. The bulk of academic studies suggest that if a tanker or terminal were hit by a rocket or had a very serious accident, the result would be a firestorm.

Writing about a possible LNG tanker fire occurring at or near Boston's Everett LNG terminal, James Fay, an emeritus professor from MIT, wrote that "the fire that would ensue would be of unprecedented size and intensity. At any point along the inner harbor route of ship travel from sea to berth, pool thermal fire radiation that can burn and even kill exposed humans, and ignite combustible buildings, will be experienced along and inland from the waterfront."[42] However, the U.S. Department of Energy (DOE) strongly disputes this possibility and points to a study by Quest Consultants Inc., which says that the "public would have little to fear from any LNG tanker spills or resulting fires."[43] The DOE claims that the study was done independently, but officials at the company that did the study say they "produced the study at the DOE's request."[44] Clearly, somebody is not telling the truth. The consultants who produced the study, also went on to say that their work was done at short notice and that the DOE was now using the study findings in ways that were not "appropriate" for the kinds of computations used. Quest claims that they had expected their study to be used as one of a number of tools for estimating LNG fire scenarios.[45]

This amounts to the DOE knowingly misleading the public. The charge is an extremely serious one. If an LNG tanker should ever explode in U.S. waters and the terminal has been built too close to a populated area, many fatalities will result. What is more, this same study, which has not been peer-reviewed, has been quoted over and over again, both by the DOE and by corporations trying to site LNG terminals in the United States. Shell, for instance, used it in their failed bid to get an LNG terminal built off Vallejo, California.[46] Intriguingly, Shell stated that the Quest calculations "were developed for the DOE in 2001." This must bring into question the value that corporations and government place on human life, versus the need to avoid the economic disaster of a carbon chasm.

The LNG Cavalry Will Arrive Too Late

There is yet another huge irony in the new rush to LNG. As Federal Reserve chairman Alan Greenspan hinted, it won't solve America's problems anytime before 2006, and possibly later, because it takes three to seven years to permit and build an LNG terminal. The United States faces the very real prospect that it might commission and start building many terminals, only to find that a profound economic downturn sets in, ironically induced by the very shortage of gas that the new infrastructure is belatedly trying to fill. Such a downturn is a sure-fire way of causing a major reduction gas demand.[47]

LNG Terminal Problems

Responding to the global awareness that gas consumption is rising past the ability of most countries to produce enough domestically, new LNG terminals are being built around the world. In the United States, there are requests to build more than two dozen new LNG facilities, and the number keeps growing. The U.S. government accepts that it is not likely that all of these will be built, because of permitting difficulties and delays, and possibly owing to financing problems if there is a global dollar crisis. However, the Houston international energy banker Matt Simmons, who has done some of the most thorough studies of U.S. gas supply and demand, has calculated that up to forty LNG terminals would be needed by 2010 to fulfill the 1999 National Petroleum Council (NPC) predictions.[48] Given that there is already enormous pressure to build new LNG terminals, close and continuing examination of the reasons for and against building them should be a high priority for everyone in a country requiring LNG imports.

The problems with LNG terminals are that, as the Skikda explosion shows, they present major safety and security risks similar to, and indeed worse than, those of tankers; they have serious environmental drawbacks; and they are also large and ugly (though this last-named is not a unique concern in a high-energy industrial system!). Those in the industry have long feared that one large accident with a vessel near a port or with the terminal itself may deal a death blow to LNG. The Hindenburg fire in 1937 ended the career of passenger hydrogen balloons. Three Mile Island and Chernobyl have slowed down or stopped civil nuclear-power reactor building in most countries. The full ramifications of the LNG accident in Skikda cannot be known at the time of writing. The accident has not gained the same public attention as the other incidents, in part, perhaps, because LNG is not yet an issue for most people in North America.

However, recent popular opposition in California, Alabama, and Mexico has shown that almost nobody wants an LNG terminal in their backyard, though the employment opportunities will be welcomed in places faring badly in the economic downturn that has begun in the twenty-first century. Corporations know only too well that nuclear accidents in the 1970s stopped U.S. civil nuclear development dead for many years. Still in 2003, despite the prospect of generous new taxpayer subsidies, the U.S. nuclear industry remained moribund.

If the siting of new LNG terminals raises awkward problems in the United States, one solution favored by the U.S. government and corporations is to site the LNG terminal conveniently in Mexico, where there are few safeguards or regulations. This would allow LNG tankers to unload in Baja California, generate the electricity in Mexico, and send the electricity across the border to the hungry air conditioners and industries of California. The risk of LNG problems and pollution associated with the power generation is neatly offloaded onto the Mexican population.

At the beginning of 2004, in the entire world there were 150 LNG tankers, with a combined capacity to transport roughly half a trillion cubic feet[49] (the United States uses that much gas in about a week). For thirty years, all of the world's LNG tankers have been built in Japan or Korea, the world's two primary users of LNG.[50] The building capacity of the six main shipyards in these two countries is about twelve ships a year, which could be increased to twenty if shipyards in Europe and China join the effort.[51] In February 2004, worldwide, there were fifty-four tankers under construction and another twenty-five on order.[52] A moderately optimistic estimate would suggest that if no ships are retired, there may be 250 LNG tankers available by 2010, though by then, some tankers may be larger than present-day vessels.

There is another highly significant difference between LNG and oil. Owing to the enormous infrastructure investments needed for LNG, producing countries and corporations generally require very long contracts, up to thirty years in fact. The U.S. markets, on the other hand, in both oil and gas, are structured for short-term contracts. Though time frames are expected to shorten, and spot markets to develop, this timing mismatch is extremely serious and will put the United States at considerable global competitive disadvantage, since other nations are able to adopt longer-term, planned strategies.

LNG—Time to Say No to Energy Addiction?

The United States will find the world of liquefied natural gas potentially much more troubling than that of oil. To the extent that the so-called War on Terror is a cover for increasingly desperate moves to control the world's dwindling oil supply, expansion into LNG (with its main production sources in politically anti-American states) threatens an even greater likelihood of endless war, covert disruption and forced regime change.

For all of these reasons all citizens of the United States should think very carefully before spending hundreds of billions of dollars and entering the global LNG market, one which may only last two or three decades anyway. After the easy natural gas has been used up there will be absolutely no other fluid hydrocarbon to turn to. The present peak and decline in North American natural gas production is the last warning that the United States and Canada—and possibly the world— will receive. We must not ignore it this time—although sadly history suggests that we shall do just that.

4

Demanding Gas

Demand—Or Why We Use So Much Gas

Given that natural gas is so useful, it may seem surprising that we came to it so late in the hydrocarbon age. However, there is a kind of natural progression from using solids, then liquids, then gases. The solids, in the form of coal, were easiest to find and easiest to handle, at least in smaller quantities. However, oil proved to be optimal in that it is, or was, available in huge and ever-expanding quantities and was also easy to handle and transport. Natural gas can't be "handled" and, being lighter than air, tends to disappear upward as soon as its containment is ruptured—unlike oil, which infamously tends to stick to the surface of land, sea, and almost anything else. Finally, as discussed in chapter 3, the difficulties of conveying gas long distances meant that a network of pipeline systems was necessary to bring gas into widespread use.

Once the infrastructure for natural gas use was well established, more and more uses were found for it. As the demand for clean heating and electrical power has rapidly increased, pushed by growing concerns about acid rain, polluted air, climate change, and the troubled safety and economic record of nuclear power, so has the demand for gas. Natural gas has seemed to many, including a large number of environmental groups, a cure-all for a lot of energy ills. Attempts to implement the weak and highly flawed Kyoto Protocol often rely on using gas instead of coal or oil. Indeed, with the obvious political difficulties of oil supply, and the well-concealed geological problems of oil,[1] the consumption of gas is being driven ever upward. In 2003, hydrogen received a large boost from the George W. Bush presidency as the "freedom fuel" that would deliver the United States from foreign energy dependency. What the president, along with most political and other analysts, failed to mention was that almost all hydrogen is currently produced from natural gas.

Energy Itself

It is reasonable to ask what energy is, and why it is so important. At the most basic level, energy and matter are equivalent, as Einstein's famous equation showed.[2] Except when considering nuclear power, however, this is not the most helpful way to understand energy. Energy is really life. All life would cease immediately without energy. Energy is the ability to do work, to move people and objects around, and to heat and cool things.[3] The essence of energy from a human point of view is mobility, from the atoms that move faster when they are hot, to all the things and people that we now move around the world in such a frenetic hurry. Without energy, the planet itself would stop spinning, so it is literally more truthful to say that energy makes the world go around, rather than money. In truth, even the economy itself is also really the flow of energy. The day is coming when we shall find this out, since we cannot eat money or use it as a substitute for energy, except to burn the paper variety.

In one sense, as far as life on Earth is concerned, energy is limitless, in that our daily source of energy, the sun, will very likely go on shining for many millions, perhaps billions, of years into the future. For all practical, even evolutionary and geological purposes, that is eternity. But limitless time cannot be translated into limitless daily motion. It is patently obvious, but usually ignored, that only so much new energy arrives on the planet each day. This means that unless we find some way of capturing more of the sun's past energy, usually by cutting down trees or mining the planet for energy stores such as coal, oil, and gas, we have to manage on that daily solar ration. The modern industrial machine age clearly doesn't believe this. Almost the whole existence of what Westerners take to be normal life is based on a massive withdrawal from limited historical energy—a withdrawal that is growing by the day and by the decade. At its simplest we are spending far more than we earn in energy terms. To help cover this unpleasant fact up, we are all operating under a system of false accounting, equally as dangerous and ultimately as doomed as any practiced by the global corporations. If we go on doing it for long enough, we will be caught. It looks like now the moment of apprehension by the laws of nature has arrived, for oil globally and natural gas in North America.

To understand how we might manage with less gas rather than more in the future, it is important to understand how we have come to be using so much of it in the first place.

Natural gas is increasingly the fuel of choice for power generators, governments, and environmental campaigners alike. Natural gas produces less carbon dioxide for a given amount of energy released and is, in many ways, the perfect fuel for power stations. In addition to producing much more carbon dioxide, burning coal, unless it is expensively cleaned, also emits acid-rain-producing sulfur chemicals, many smog-inducing particulates, and mercury, which is causing growing concern to many governments. Burning oil and its derivatives may be considered somewhere in between gas and coal, but the awareness is dawning that the world's oil reserves are not growing, and that oil supply is in trouble.[4]

Kyoto Needs Gas

"Kyoto" is often mentioned in the same breath as natural gas. Although the Kyoto Protocol will almost certainly be seen by history as a great and perhaps deliberate distraction by industry and governments who dare not face up to the real measures necessary to reduce climate change, namely, economic contraction and reducing population, nonetheless it is important to understand what Kyoto is, at least in outline. In 1997 United Nations delegates agreed to legally binding targets of an overall 5 percent reduction on 1990 emission levels for six of the main greenhouse gases.[5] The target date for compliance is between 2008 and 2010.[6] At least fifty-five countries accounting for more than 55 percent of emissions would be required to ratify the agreement to bring it into force.

In 2001 the United States declared that it would definitely not ratify Kyoto. Australia followed. Canada wavered for a long time, but following the granting of extraordinary loopholes at the Bonn meeting in 2001,[7] it finally signed on in 2002. There are suggestions that it did so because outgoing Prime Minister Jean Chrétien was trying to clean up his poor record on the environment, with the near certain knowledge that the Protocol would never in fact come into force. Chrétien's successor, Paul Martin, seems to be vaguely in favor of Kyoto but has no specific plan for how to implement it.[8] By the end of 2003, six years after the initial parties signed in Kyoto, the protocol was still not ratified. Russia was the last hope for making the 55 percent a reality. But in October of 2003 and again in January 2004, a top economic adviser to President Putin made it clear that Russia was highly unlikely to ratify Kyoto.[9] Pulling in the other direction, the European Union suggested that Russia might be offered easier entry into the World Trade Organization (WTO) if it signed Kyoto.[10] However, some E.U. states are now suggesting that if Russia doesn't sign, it may have to weaken its own stance on greenhouse gas emissions or risk being less competitive.

In a sense, by this time, Kyoto was already irrelevant since there was no conceivable way that the targets could be reached globally. Even Europe, chief proponent of Kyoto, was set to miss its own self-imposed targets, and the only reason there was success in reducing European emissions during the 1990s was accidental anyway.[11] One of the two countries helping that reduction was Britain through its policy of converting rapidly to gas in the 1980s and 1990s. This had nothing to do with averting greenhouse emissions, though Margaret Thatcher was one of the first world leaders publicly to recognize the problem in 1988. Rather, it was for traditional political and economic reasons in that Thatcher wanted to punish the coal miners for unseating the Conservative government in 1974,[12] and Britain temporarily had abundant and cheap domestic natural gas from the North Sea. That abundance is predicted to disappear dramatically after 2005.[13]

Britain provides a good demonstration of the main way countries are hoping to reduce their greenhouse emissions, whether to satisfy Kyoto or not. Converting to natural gas is an economically painless way of carrying on business as usual. It means no change at all in energy usage and indeed often encourages more consumption, via the increased economic activity associated with the technological approach to "greening" the economy. This shift to gas is a mistake that will have immense consequences. Kyoto will, ironically, have helped push the world further into carbon addiction, even though it may have some symbolic use in setting legally binding limits. The fact that almost every country is going to fail utterly in meeting these limits should cause us to question the way these kinds of global agreements are developed.

No Limits to Substitution

Energy is not unlimited, but we believe it is because our way of life simply depends on energy being unlimited. Strongly reinforcing the illusion of "no limits," we have a highly developed economic system that relies on "substitutability." When cod disappear from their largest habitat off the coast of eastern Canada, we can eat some other fish, like orange ruffy, until that too disappears. When ocean fish have so declined that there are calls to ban fishing from one-third of the oceans,[14] the economic answer is to do more fish farming, what some call the "Blue Revolution."[15] There is apparently always something new we can substitute for what we have used up or destroyed. So it is assumed with energy. In Britain, in the seventeenth century, having cut down almost all the abundant forests, people began burning the dirty black lumps they found on the beaches—sea coal. Although coal was dirty and burned with poisonous gases, it had a higher energy density and a higher energy profit than wood. For a given effort in finding it, it allowed more work to be done. Coal was soon found in vast quantities in northern England and Scotland. Thus coal was a good energy substitute and was the first fuel of the Industrial Revolution. Production in Britain rose dramatically, doubling every fifty years throughout the eighteenth century, and rising by a factor of 25 during the nineteenth century to reach a quarter of a billion tons by 1900, almost a hundred times what it was in 1700.

Simultaneously with the increasing coal production in the eighteenth century, classical economics was being developed, and not surprisingly energy doesn't enter into it at all, any more than the availability of gossip or sea water. This historical context of ignorance[16] and abundance is at least one reason why our whole industrial system has particular trouble understanding that it is based on energy, and that energy sources may one day be insufficient. Problems of energy form no part of traditional economics, and even if a problem should arise, energy prices would rise, and substitution would solve the problem. That is the theory. Sometimes substitution happens even when there is no problem with availability or price. In the middle of the nineteenth century, commercially recoverable quantities of oil were discovered in the United States (though oil had already been in use for centuries in Russia). Hoewever, it was not until 1901 in Texas that the first giant field was found at Spindletop, near Beaumont. So began the American century, which was really the oil century. Oil was an excellent substitute for coal and in many ways is a much better energy source. It comes out much faster than coal, doesn't require men to go underground and dig it, and has a much higher energy pay back than coal.[17] Oil is extremely energy dense—that is, it contains a lot of energy for its weight and volume. It is also very easy to transport and use in industrial processes, which coal is not. Oil was a welcome substitute by choice; the United States already had the world's largest coal reserves and still does.

The 1901 Texas oil and gas find was the beginning of over sixty years of discovering more and more large oil fields, the largest being Ghawar[18] in Saudi Arabia. After Texas finds in the 1930s of fields much larger than the Spindletop giant, fears of an oil shortage in the 1920s quickly changed to demands to keep production down and prices up. For example, after the Spindletop find, prices dropped to 34¢ a barrel. But in the 1930s, following the new discoveries in Texas and other states, the price was pushed below 10¢ a barrel. That's just over 1¢ a gallon, or a quarter of a eurocent per liter. Little wonder that people thought energy was free. But even then energy wasn't free, and the producers demanded protection and controls. In fact, though OPEC is vilified by many, it was modeled on the Texas Railroad Commission, which controlled the U.S. oil and gas market for most of the first half of the twentieth century.

Apparently limitless and expanding quantities of cheap and easy oil immensely aided the growth of global industrialization, in both the capitalist and communist worlds. In the process, industrial society became almost totally dependent on oil, and for more than a century ever more oil was found each decade across the planet until the mid-1960s. However, by 1981, the discovery rate had declined so much that we started using more oil than we found. Whether the inevitable peak of production has been reached is still being hotly debated, even among those who accept the idea that oil extraction will soon be in decline.[19] By contrast, governments, publicly at least, seem to believe that oil supply will go on increasing for decades. Remarkably few petroleum geologists believe this, but most petroleum corporations fear to utter the "D" word, depletion.

As a final comparison with coal, which took 200 years to increase 100-fold in production, oil took 100 years to increase 200-fold. The phenomenal growth of industry in general is based on the growth of energy. The same is true for the explosive growth of the human population. These two factors should be kept in mind throughout the rest of this book.

For the industrial world, limitless availability of energy is now a matter of faith, born of the experience of plenty and convenient ignorance of global reserves.

Gas Gets Its Start

In chapter 3, the early development of gas pipelines was described. The main use then was for lighting, and the way that need developed offers important lessons for our present situation. In 1792, William Murdock, a British inventor, worked out a system for putting gas to use on a broad scale by devising a lighting system for buildings. Murdock used the gaslight system to brighten up murky English cotton mills, which made it possible for people to work longer hours, and thus increase both productivity for the factory owner, and no doubt, the misery of the workers. It also meant that the factory would need to consume more energy, most likely coal, in this case.

Subsequently, Albert Windsor, a German businessman living in England, used Murdock's basic idea to design and obtain a patent for a streetlighting system. Before Windsor's system gained wide use, people traveled dark streets carrying their own lanterns. Delivery systems for the gas were organized, resulting in the development of the first documented gas company formed in England in 1812.[20] Having safer streets is clearly a good thing. But it then stimulates more nighttime activity, which usually increases economic activity in general. Thus, just as with lighting factories, the availability of more light has a compounding and symbiotic effect on all human activity, so that we want to use even more energy. The growth can easily become exponential, as long as there is enough new energy to fuel it.

Nevertheless, natural gas was obviously good for lighting and certainly burned with fewer toxins than gas made from coal. However, gas still had to piped and lit every night. Finally, toward the end of the nineteenth century, electricity took over from flame lighting, and gas producers began to look for other ways to use their product.

Electricity

Ironically, even though electricity at first displaced natural gas use, electricity is now one of the primary drivers of increased gas consumption. Because gas is so versatile, it can be used to provide the continuous baseload electricity that is needed all the time, but also to provide peaking power on hot summer days and at other times when demand is suddenly high. The current vogue for increasing the amount of electricity made with natural gas is highly significant, because electricity is now vital to human existence across the planet, especially in cities. Electricity is used for everything, including pumping water to houses, businesses, and factories, and processing the sewage waste that leaves them. This applies equally to less-consuming countries. One of the major grievances of Iraqis after the official end of the 2003 Gulf War was that they had no electricity, and hence no cooling, no water delivery, and no sewage processing. We live in an electric age, and thus anything that threatens the electricity supply is a direct threat to the lives of billions of humans. So although natural gas may seem unrelated to the electricity user, problems with it are not.

When an economy grows, its use of electricity grows. The relationship is not exact, and obviously if a place, such as California, tries to use electricity more efficiently, then its use will not increase as quickly but rise it generally will. In the United States, a 2.5 percent rise in economic growth requires just over a 1.5 percent increase in electricity production.[21] In less industrialized countries, the relationship is similar, although the rate of economic growth in some cases is much faster. Global use of electricity is predicted to double from 2000 to 2030.[22] Time will tell whether this comes to pass, but this growth has certain grim implications.

Electricity, at the scale required for industrial society, is currently generated either by coal, nuclear, hydro, or natural gas, with coal supplying anywhere between half and three quarters in most countries. In the United States, coal supplies half, and natural gas supplies about one-fifth of electricity production. Coal's well known drawbacks have made it an obvious target for replacement by natural gas, which is more convenient for making electricity and much cleaner. In the United States, the fall from grace of nuclear power over twenty years ago has left the field clear for natural gas. That, combined with its low cost during the 1980s and 90s, has made it the fuel of choice for new power stations and indeed encouraged many older stations to switch from coal. By the end of 2003, over 300 new gas-fired power stations had been built in the Unites States. When these stations begin to reach high levels of utilization they will have a dramatic effect on gas demand.[23]

Despite spending $100 billion on this building program, it seems that no one in the industry or government had stopped to ask where all the gas would come from.

Or had they? It appears that many power generators did do some homework on North American natural gas supply, and were evidently satisfactorily reassured by the 1999 National Petroleum Council (NPC) report. The three-volume report, titled "Meeting the Challenges of the Nation's Growing Natural Gas Demand," was, according to energy analyst Andrew Weissman,

> one of the few government or privately-sponsored studies that offered any substantial basis for believing that it would be possible to significantly expand North American supplies of natural gas above 1999 levels. It also provided the basis for many of the assumptions used by the Energy Information Administration (EIA) in its subsequent annual forecasts of supply and demand in the U.S. market.
>
> As such, it played an important role in justifying decisions by power plant developers to build more than $100 billion in new gas-fired generating units over the past four years—foregoing the opportunity to construct a more diversified portfolio that relied more heavily on coal-fired generation and renewable energy.[24]

Such a claim is so extraordinary and devastating that if proven it should provoke a national outcry. In great part, this colossal rash of power station building has cost the United States precious time in trying to adjust to a landscape that will be seriously short of natural gas. The 1999 NPC report is in its way, however, just one more example of the kind of cornucopian delusion that characterizes many in government and most in industry, who believe in nothing but economics and the miracle of capitalism with its unlimited ability to find substitutes for everything.[25] Human ingenuity can be harnessed to find a solution for every problem, so the thinking goes—as long as there is a market to pay for it. This belief system is bringing the planet to the point of total ruin. If merely naïve, it is a wretched testament to human greed, myopia, and selfishness; if not, it is criminal deception. Either way, the chief conclusion for the walking worried, the concerned citizen, should be that the public cannot and should not trust the leaders of government or corporations to tell the truth about energy.

Europe

Europe, too, has caught the gas bug, especially since it ratified the Kyoto protocol. Britain, having discovered large gas deposits in the North Sea in the late 1960s, indulged in what has been called the "dash for gas." It has become dependent on a fuel now in sharp decline from its own fields, which means it will likely be importing one-third of its natural gas by 2010.[26] Britain has confirmed that it will certainly start importing gas in 2005.[27] Only France and Lithuania have low dependence on natural gas for electricity generation, and that is because more than three-quarters of their power stations are nuclear. However, these systems are old, and in the summer of 2003, France discovered they were also vulnerable to low river levels, which reduced their cooling water. France has been debating the huge decision of whether to replace its aging fleet of nuclear power stations, or move away from nuclear, because their fast breeder program has failed, and they have no real plan for dealing with the waste.[28]

Thus, the whole industrial world is now looking to natural gas for its electrical power, but no country in this group, with the exception of Norway and Australia, will have enough domestic gas by 2010. The rest will be competing to import gas from the same places. The only other possible exception is Canada, but to have enough gas for its own uses, it would have to terminate the energy chapter of North America Free Trade Agreement (NAFTA), an unlikely event, especially as Canada's current prime minister, Paul Martin, seems to have a strong desire to maintain close ties with Washington.

Home Uses

Natural gas has become very popular for home (and office) heating of every kind, including cooking, making hot water for bathing and showering, and what is called "space heating." Though it may sound like it, space heating is not a boondoggle, make-work project from NASA but simply refers to heating our living space, either by "forced air" in much of North America, or centrally heated hot water radiators[29] in Europe. More than 60 percent of U.S. residences are heated with gas, as are 70 percent of new buildings, and the figures are higher still in Canada. The percentages in Europe are similar or higher. Across the world, as usual, other countries are following the Anglo-Western lead and converting their heating to natural gas.

The issue of heating is one of the most serious issues concerning natural gas. Once a single-fuel heating system of any kind is installed, it is very hard, and very expensive, to change it. Most people won't do it and will either pay for higher gas prices, possibly at the expense of other vital items, or else will do without heat, and be cold. In many parts of North America, especially in Canada, being cold can mean freezing to death. This will obviously affect the poor and the elderly (often the same group) much more than the rich. The insulation, energy-saving and generating measures, which could have been introduced over the last thirty years since the oil shock and the mid-70s U.S. gas crisis, have generally not materialized or have even been eradicated where they existed.

Cooking with gas has many advantages, since it is both more responsive and easier to control than electricity. Many restaurants rely on gas, particularly for the large ovens required for pizzas and deep fat fryers. Since one in four meals in North America is eaten in a restaurant, this is no small matter. In less industrialized countries, gas cooking produces much less greenhouse pollution and waste toxins than burning dung or biomass such as wood.

Fertilizer

Although electricity generation and heating are absolutely vital uses of natural gas, there is an even more significant and important use of natural gas, which now affects most of the planet's human population, even if it is largely unknown—the production of industrial agricultural fertilizer. If asked to name the most important event associated with petroleum, most people might think of 1886 and the invention in Germany of the horseless carriage, or car, powered by an internal combustion engine. Very few, if any, would think of 1909 and the artificial production of ammonia, one of the three key ingredients of industrial fertilizer, which provides the extra nitrogen that modern agriculture requires in such vast quantity. Almost all ammonia is made from natural gas, using the Haber-Bosch process.[30] It was the first economical method of synthesizing ammonia from nitrogen in the air and hydrogen and is still the main method in use today. The process occurs at high pressure and at considerable temperature, but it is the supply of hydrogen from natural gas that is the critical part, since the atmosphere contains only very low concentrations of hydrogen, despite its being the most abundant atom in the universe.

It is possible and reasonable to argue that the staggering rise in the human population during the twentieth century owes more to the use of natural gas to make nitrogen fertilizer than any other single factor, even medicine.[31] More than 2 billion humans, mostly in the poorest countries, depend directly on artificial nitrogen fertilizer for their daily food. This obviously benefits the small number of global corporations that control the business, since it gives them enormous power and control over the lives of billions of people and guarantees their profits. For now, the global supply of natural gas is still increasing so that in theory there will be no nitrogen fertilizer crisis for some time.

Unless, that is, the situation in North America continues to decline. Therefore, it is worth documenting the case of the United States, even though it is too early to say how matters will be resolved. Furthermore, the United States is a paradigm, since it is both very large and very influential.

It is both interesting and alarming to record that by 2003, half of the nitrogen fertilizer industry in the United States was lying idle, and, according to industry representatives, one-fifth had had shut down permanently. There are two obvious and important questions arising from this. Why did the fertilizer industry get hit so hard, and where is the rest of the necessary fertilizer coming from?[32]

It seems that the fertilizer industry had borne the brunt of the three years since 2000 of what market analysts call "demand destruction," a strange euphemism for economic contraction. Some 80 to 90 percent of the price of fertilizer is natural gas, which means that producers are highly sensitive to rising gas prices. At around $3.50 per 1,000 cubic feet the industry says it is no longer economical to manufacture nitrogen fertilizer in the United States. The price has rarely fallen below this figure since 2000,[33] which explains why the industry has declined so much during the last three years. The drastic and politically highly sensitive refilling of the gas storage system in 2003 (ahead of the 2004 election) was being done mainly at the expense of the fertilizer industry. By the time of the official U.S. government natural-gas crisis conference in mid-2003, the American industry was wondering how much more "demand destruction" it would have to withstand.[34]

But the sudden reduction of U.S. nitrogen fertilizer production leaves a continuing and possibly worse problem: North America is still the export breadbasket of the world, Canada and the United States being the world's largest exporters of grain. The soils of much of the once-fertile corn and wheat belts are badly degraded and eroded, which means that grain production is both dependent on and addicted to artificial nutrients, mostly made from oil and natural gas. The troubled soils either need more fertilizer to prop them up, or else a massive program of rolling back the "Green Revolution"[35] that helped ruin them in the first place. It will be no surprise that a "rolling back," involving reverting to nonpetrochemical food production, may never be on the orthodox economic menu. Without such a dramatic change, however, the United States will be forced to import more and more fertilizer. The recipe may sound familiar.

In fact, there are several countries very willing to produce fertilizer and sell it to the United States, because they have what is called "stranded gas," or gas too far from any industrial settlement to be economically conveyed there by pipeline. In particular, Saudi Arabia has large reserves of gas that it has never sold in quantity and is now happy to convert into fertilizer. The fertilizer sells for more than the natural gas, and in a particularly felicitous stroke, this means that the water-starved Saudis can export some of their drought problem too, since the fertilizer is used to produce grain in the United States, with American water. The Saudis then import the grain at a very reasonable price. Of course this does nothing to help the looming water crisis in the United States.[36]

Not only is natural gas vital for fertilizer; in modern agriculture, natural gas may be used to power water irrigation pumps and is heavily used to dry many products, including corn.[37] With the arrival of higher natural gas prices in 2000, North American farmers have increasingly had to engage in complex economic calculations, balancing costs against risks associated with the weather and letting their crop stay longer in the field before harvesting it.[38] With luck, delaying the harvest saves natural gas because the crop requires less drying. The risk is that more crop will be lost than money saved by not using the gas. The situation is compounded by the problems of farmers planting less, again because of more expensive fertilizer. North American grain harvests have been falling for several years, and the problems with natural gas are only exacerbating the issue.

Industrial Uses

Most plastic is made from either oil or natural gas or both. Some chemical manufacturers are starting to accept what the petroleum industry refuses to admit (at least openly), that the beginning of the twenty-first century marks the end of the age of plentiful and cheap oil, and they are increasingly looking at natural gas as their main feedstock. Evidence of this is found in new processes to make more plastic from natural gas, via methanol[39] and the direct production of acetylene from natural gas.[40] The U.S. situation is different from most of the rest of the world, since its plastics feedstock industry developed at roughly the same time as the increasingly abundant supply of cheap natural gas, mainly in Louisiana and Texas. This is where about half of the industry is located. When the gas price started to climb fast at the beginning of the century, these industries were hit hard. By 2001, a Houston economic journal estimated that half of the methanol industry and a third of the ammonia-producing industries had shut down, thanks to natural prices more than doubling. They pinned the blame on weather and the rise of natural gas power stations.[41]

Before the end of the nineteenth century in places such as Kansas that had ready access to natural gas, it had become a popular source of cheap energy for industrial uses such as brick plants, zinc smelters, and cement plants.[42]

Transportation

Certain types of transport, especially buses, are seeing an increase in the use of natural gas. This usage is a small percentage compared with heating and electricity, but it is growing. Natural gas is cited as burning cleaner than gasoline or diesel, with less nitrogen oxides and particulate matter.[43] Particulates are small pieces of solid or liquid matter such as soot, dust, fumes, mists, or aerosols. They have a complex and sometimes moderating effect on global climate. They also defeat the body's natural clearing mechanisms and are often chemically active and acidic, which only increases their damaging effects on health.[44] Various large numbers are cited for the death toll from particulate pollution, both in the industrialized West, and even more so in the industrializing East. It may be safest simply to say that producing less fossil-fuel derived particulates would be a very good thing for all life forms.

Advocates of natural gas vehicles can point to the fact that such vehicles do emit less pollution of all sorts, including less carbon dioxide. Another reason for their adoption is that they could be a transition to using hydrogen fuel-cells in vehicles. The primary source of hydrogen as fuel or feedstock is natural gas.

By 2003, there were estimated to be more than 70,000 natural gas vehicles (NGVs) on the road in the United States, and their numbers are growing all the time. Passenger vehicles have a lifetime of over ten years, and public transport vehicles are typically double this. Therefore, public transport systems are much slower to change than private cars, and decisions about them should be based on a longer time frame. Since North America is facing increasingly severe natural gas supply problems, it is clearly unwise and will likely prove a very poor investment to convert buses to natural gas. More widely, converting to natural gas as a transport fuel will only increase the demand for global natural gas and place further political and possibly military pressure on the few large producer countries.

Environmental Concerns

Natural gas is called "clean" because it burns to produce fewer toxic pollutants and less carbon dioxide for a given amount of energy released. While this much is true, it is not the whole story. Methane (the principal component of natural gas) is itself a very serious greenhouse gas, over twenty times more powerful than carbon dioxide. It is calculated that over the last two centuries its concentration in the atmosphere has doubled. Even the U.S. government attributes this to human activity.[45] Overall, of the gases produced by industrial society, methane is considered to be second only to carbon dioxide in total contribution to the warming and disruption of climate. Natural gas also causes low-level ozone or smog problems, possibly more so than diesel.[46] In its favor, it endures in the atmosphere for a shorter time than carbon dioxide, roughly twelve years.[47] However, methane's high potency as a greenhouse gas and its relatively short life make it a good target for reductions, since it is thought that reducing its level would have an effect more quickly than reducing carbon dioxide. This would likely be of more interest to politicians than measures that will produce little obvious benefit during their time in office.

The problems of venting or losing natural gas are worse in less industrialized countries or countries like Russia that have more or less given up any pretense of worrying about the environment. The Russian president in 2003 said that 15 percent of Russia could be classified an "environmental disaster zone."[48]

Natural gas has been vented (and flared or burnt) for decades in oil-producing areas that have no use for the gas that so often accompanies oil extraction. Countries like Nigeria, which have done this in the past, are now starting to build facilities such as LNG terminals to use their natural gas. Nevertheless, although it is difficult to be sure of exact numbers, natural gas is still routinely vented to the air, particularly in Russia, which has one of the worst records for pollution from the oil and natural gas industry.

Natural gas escapes during processing in refineries and at the well-head. There are reports suggesting that the amount of natural gas escaping during the extraction process is much more than previously realized, possibly even double. According to U.S. government research, this would mean, for instance, that the United States is emitting between 4 and 6 million tons more methane per year than is presently being accounted for.[49] It was found that smog in Oklahoma City, capital of America's third largest gas-producing state, was worse than in Los Angeles.[50] It is highly likely that other major gas-producing areas, such as the Middle East, Venezuela, and Russia, have also underestimated the amount of methane escaping, thus further complicating calculations on the greenhouse effect, as well as worsening the problem itself.

Sour Gas

Another problem with natural gas, which is seldom mentioned by its proponents, is that much of it is "sour." Sour natural gas contains significant concentrations of highly toxic and carcinogenic hydrogen sulfide, usually more than 1 percent,[51] however, the amount may be as high as 5 percent. Many of the world's larger and newer important gas producers suffer from this problem. Northern Iraq has famously sour gas, and so does much of the Caspian basin, including the big reserves in the north. For sour gas to become useable, the hydrogen sulfide must be removed. In some places the resultant sulfur is a useful by-product; in others it produces dangerous mountains of waste. Sour gas from wells has been released into the air, killing gas workers and causing human and animal health hazards in the locale, even in "first" world places such as Alberta, 30 percent of whose gas production is reckoned to be sour.[52] As a result, in Alberta, there have been various cases of sabotage against gas companies to stop them drilling sour gas wells.[53]

Human Rights

In another tragic sense, natural gas is no cleaner than oil. The same practices of condoning human rights abuses and environmental destruction that have attended oil production throughout the twentieth century also apply to natural gas. Aceh, in Indonesia, is a well-known example, where ExxonMobil has been accused, based on strong evidence, of not stopping "murder, torture, and other crimes by security forces guarding its gas fields," even though, according to its accusers, it knew what was going on.[54] If petroleum corporations admitted when mistakes were made in a more open way they may not be so widely distrusted. Too often however it appears that only when great citizen effort is made, or as in the case of Shell in Nigeria, the abuses become widely publicized, is any wrongdoing accepted. Even then, these corporations will go to inordinate lengths to avoid retribution. Fifteen years after the worst tanker oil spill in history, at Prince William Sound, ExxonMobil is still trying to avoid paying punitive damages.[55]

The phenomenal rise in gas demand looks set to accelerate and encompass the whole world. Gas is certainly being globalized. But just as globalization is a harbinger of problems in every other sphere, so it will be with natural gas. Those enacting plans for the global dash for gas are making the same mistake as the U.S. power producers did in 1999. They believe that gas is unlimited. It's not, and the unwelcome specter of depletion will lead to gas peak, just as surely as it has led to oil peak. After natural gas, however, there is no new hydrocarbon savior, only a return to old coal and nuclear power. Unless of course, the industrial world has somehow woken up to limits to growth and begun both a massive shift to truly renewable energy and a contraction of the world economy.[56]

5

Gas Ability

Discovery Peak

You cannot extract what you have not discovered is an old truism among petroleum geologists. Veteran exploration geologist Jean Laherrère has taken this idea and turned it into a series of charts that compare the rate of discovery with rate of production. They tell a powerful story, which graphically illustrates the limits of fossil energy.

Although there is much disagreement among authorities, a reasonable figure for the amount of ultimate recoverable conventional gas in the world is 10 Peta cubic feet[1] (Pcf) or 10,000 Tera cubic feet, with an additional 2.5 Pcf of unconventional gas.[2] In 2001 about 90 Tcf of gas (both conventional and unconventional) were produced and marketed.[3]

In figure 5.1 the thin discovery line shows the amount of gas discovered globally reached a very sharp peak in 1970 when the huge field in Iran and Qatar[4] was found, then fell off equally quickly. There was a small recovery at the end of the millennium, but in the years 2001 and 2002, for the first time, less gas was discovered than used. This is

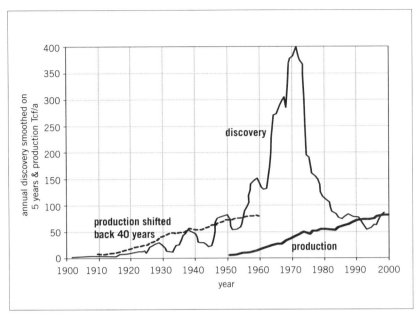

Figure 5.1. World conventional natural gas annual discovery and production.
This chart shows Laherrère's graph of global natural gas discovery and production. Production is also shown shifted back forty years, to illustrate how it follows the pattern of discovery. Note that annual data tend to be erratic so smoothing is done by taking the average of say two years before and after a given date (one can add more than two years either side, as some other graphs in this book do). This gives a more useful result, but it does mean that the first and the last years have to be removed as incomplete. In figure 5.1, for instance, this explains why the graph stops before the year 2000, even though there are data for 2000 and 2001. *Source: Jean Laherrère*

highly significant: twenty years after the same thing happened to global oil (in 1981[5]) the world entered the oil peak-plateau.

The thick line shows how the world has increased its production to date. However, if you take that line and move it back forty years on the graph (the dotted line) one can see that gas extraction has so far followed the discovery pattern (just like oil before it). If extraction continues to follow the general contour of reported discovery, a colossal rise in future extraction would reach an extraordinary peak in 2010 and then drop like a stone. In reality, this is extremely unlikely for a number of reasons, so production will be greatly smoothed out. Since we can only use what we find, if world reserves are over reported or new fields of natural gas keep failing to replace the amount consumed, then the question is for how long will production (the thick line) keep rising to meet the world's increasing demand?

Furthermore, the amount of infrastructure required to increase gas extraction at the rate to follow the discovery line cannot be built fast enough.[6] Because much of the new natural gas is remote, or "far from market," a vast matrix of pipelines, much of it serving a colossal new global LNG system, will be necessary to realize the "globalized" gas market now envisaged. This infrastructure will be on a scale so far unmatched on the planet. Though technologically feasible, the sheer scale of these pipelines, along with the LNG tankers and processing plants, may yet stretch the world's financial system past its capacity. With the United States facing unparalleled financial problems, which may devastate the world's financial system, it is possible that the enormous investments required will not be forthcoming.[7] There may also be problems with nickel shortages,[8] which could slow down construction of the huge, high-quality steel pipes required for transportation.

These limitations will likely flatten the global gas extraction pattern considerably, though fast rising demand will try to pull production up as quickly as possible. A gas peak-plateau somewhere between 2010 and 2025 seems within the realm of possibility. Three things would change this forecast considerably: there are sudden and new gas finds of immense magnitude; ways of safely and commercially tapping methane hydrate deposits are found; there is an enormous global economic crash. The last item seems by far the most likely. This would certainly stave off a gas decline for a number of years, possibly even a decade or two, depending on its severity. Followers of the "Olduvai Theory of Industrial Civilization"[9] or "Die Off"[10] devotees might argue that the world economy will not recover from an oil-peak-induced crash. These are obviously highly speculative issues, but the stakes are so high, that speculate one must or else give up on all long-term policy and planning other than the cornucopian, business-as-usual type.

Any consideration of gas or oil extraction should include at least a brief mention of Marion King Hubbert, who carried out the seminal research on oil production curves, and who lent his name to the Hubbert Peak. In 1956, he was chief researcher for Shell in Houston when he made his famous prediction that U.S. oil production would peak around 1970. He was ridiculed at the time, but he turned out to be a real Cassandra—telling the truth, but not being believed. It took the 1973 Arab oil embargo to hammer the message home, even though there were reports of gas lines in 1972, before the oil embargo. This was not the first warning signal to be missed. Oil and gasoline shortages appeared in 1948 in the United States, which, interestingly, was the first year in history that America imported more oil than it exported, making it a net importer.[11] In the decades that followed, U.S. foreign oil dependence became total addiction. The same is now true for natural gas.

Hubbert's Peak As It Applies to Gas

Hubbert's Peak is a curve, sometimes bell- or almost mountain-shaped, that describes the rise in extraction of a fluid, be it oil or gas, from a whole reservoir[12] of connected porous rocks, or a collection of reservoirs such as make up a nation's oil reserves. Theoretically, if engineers knew exactly where all the oil or gas was, they could drill hundreds or thousands of vertical wells (or slightly fewer horizontal wells) in just the right places and extract the reserves, much like emptying a car's fuel tank. But several factors prevent this from happening or even being possible. Oil and gas exist underground in porous rocks, which means that though they can flow, they cannot move quickly as they would in air. Granted, they are often under very high pressure, which is what makes a gusher, but that pressure is eventually lost. Thus extraction is slowed down by the composition of the reservoir. Furthermore, of course, lack of perfect knowledge, and lack of infinite technological and financial resources, prevents a rash of wells all going in at once.[13] In real life, wells are drilled serially and with great care and some patience, since no one wants to sink even a single dry hole, let alone a few dozen.

When looking at an oil province[14] or nation as opposed to just one field, or indeed one well, there is also the consideration that discovery may still be proceeding while the earlier fields are extracted. If, as often happens, the biggest fields are found a little way into the discovery and production process, then that again will reinforce a rising curve. This same combination of geology, porosity, permeability, and flow-limiting, technical and financial restraints means that production will rise only at a certain rate. The peak, or plateau, comes when the natural pressure of the well or field starts to reduce sufficiently that the permeability takes over and increasingly limits the flow. In the case of oil, drillers may choose to pump a substance such as water or natural gas back into the well to increase the pressure. Natural gas follows the same basic principle, but, being a gas, it flows much more quickly for a given porous rock structure.[15] The nature of gas also means that as much as 80 percent of the reservoir can be recovered, whereas oil recovery is about 40 percent. On the other hand, it is much harder to repressurize a gas reservoir once it passes its peak.

In general, the peak of production for an oil province happens when the system is about half depleted. It gives the producers a pretty fair idea of how much more can be extracted, but after the peak, the rate of extraction gets slower and slower, until at a certain point it is no longer economical to extract it because the production costs outweigh the value at the wellhead. By contrast, with gas, this decline can be very rapid indeed and is difficult to reverse.

One of the main limiting factors in oil production is the number of wells sunk into a field. With gas this certainly also applies, but there is an additional limiting factor, that of the pipelines into which it feeds. The gas could, of course, simply be vented or flared, but increasingly that is seen as economically foolish and environmentally unfortunate. This leads to the possibility of what some call the "gas cliff." The pipeline capacity limit effectively cuts off the top of the bell curve, so that gas output looks flat for a long time. While in some ways this is better for a producer, who sees a more certain and predictable return on investment, it also tends to hide the warning signal that an oil producer receives when a well (or reservoir) goes into decline. Decline of an oil well may occur for many years before it becomes uneconomical. With a pipeline-connected gas well, that decline, coming later in the bell curve, not the gentler shoulders of the peak, will be much steeper, often appearing more like a cliff, and may thus arrive with very little warning.

Sunset Provinces

United States and Canada

In the case of the United States as a whole, this lack of warning about gas decline has been compounded by several factors, including the rise of offshore (Gulf of Mexico) and unconventional-gas production. According to the U.S. Department of Energy (DOE), in 2003 U.S. gas reserves[16] were 183 Tcf.[17] U.S. production fell from 19.61 Tcf in 2001 to 18.96 Tcf in 2002.[18] The number of wells rose from 373,304 in 2001 to 383,626 in 2002, the largest number ever. Although the increasing number of wells is good news for drilling-rig companies, more wells producing less gas is one of the surest signs that production is peaking, and that the situation is not just a blip.

Now that the United States, as the world's largest gas user and former largest producer, is going over the gas peak,[19] if not cliff, it is

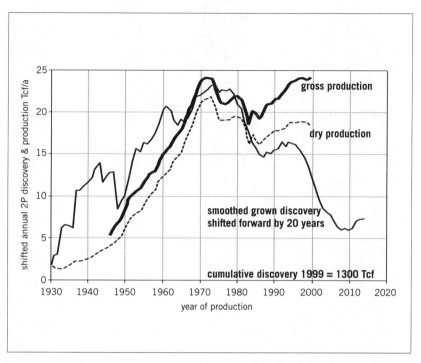

Figure 5.2. The pattern of production and (shifted) discovery of natural gas in the United States. As explained in chapter 2, the designation "2P" means "proved and probable." Gross production means all the gases that come out of the ground; dry means that which goes into the pipeline with the condensed liquids removed. The thin line is discovery, shifted forward by twenty years so that it can be easily compared with production. *Source: Jean Laherrère*

important for the American reader (and anyone else affected by the American economy) to know where gas currently comes from in the United States itself. Otherwise, it will be hard to decode the confusing and curious policy twists coming out of Washington, D.C., or to develop personal and community plans that will be somewhat insulated from U.S. official and corporate policies.

Figure 5.2 shows the discovery pattern of natural gas for the United States. Gas discoveries rose up like the Himalayas for more than half a century, then abruptly reached a heady peak in the 1950s, after which they have fallen away mightily, with just a bump in the 1970s,

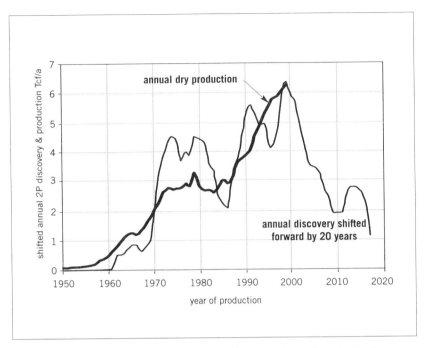

Figure 5.3. The pattern of production and (shifted) discovery of natural gas in Canada.
Similar to the United States the thin discovery line has been shifted forward by twenty years, so that the thick production line fits well with it. *Source: Jean Laherrère.*

and a small up-tick at the end of the 1990s, which has since fallen away. The major U.S. producers have now all abandoned onshore exploration. Extraction has followed the discovery pattern remarkably faithfully at a distance of about twenty years, with an all-time extraction peak in 1973, followed by a second softer—and much lower—peak in the late 1990s. The United States is now on the final downhill gas run, just as it is with oil.

In January 2004, Canada's proven natural gas reserves stood at 59.1 Tcf. Canada produced about 6.6 Tcf of natural gas in 2002.[20]

The picture for Canada, which has supplied almost all U.S. gas imports since the mid-1980s, looks, if anything, much worse than the United States. As figure 5.3 shows, following the discovery curve,

Canadian production more than doubled in the twenty years from 1980 to 2000. After a brief plateau, production has begun to fall even faster than some of the worst public predictions had suggested. In 2003 Canadian production fell by 5 percent, leading to as much as a 15 percent decline in exports to the United States in 2003, just when it was trying to make up for its own production decline. In fact, the decline of U.S. conventional onshore gas production has been masked by a large growth in coalbed-methane extraction in several areas of the United States and offshore production, particularly in the Gulf of Mexico. Without this new production, the United States would have had much earlier warning of the impending supply crisis.

As figure 5.4 shows, more than half of U.S. domestic gas production comes from the Gulf Coast, on- and offshore, with a large share coming from what is called the Outer Continental Shelf (OCS).[21] The other major area is the Rocky Mountains, which also produce a lot of unconventional gas.[22] While the governments of the United States and Canada are now publicly admitting that gas production is facing real difficulties, they won't say that production has peaked, rather that drilling levels are reduced in some places for reasons they cannot entirely explain with market theory.[23] The government numbers all confirm that production is in decline across most of the United States and Canada, with the sole exception of the relatively small gas province of the Scotian Shelf of the coast of Nova Scotia. Gas (and oil) production numbers are subject to revision, up to a year or more later, but preliminary numbers from 2003 not only confirm the picture of decline but strongly reinforce it, with declines of up to 10 percent for some of the largest U.S. and Canadian producers.

In future forecasts of gas production, much is made of the Arctic regions of Alaska and the Mackenzie Delta. The Mackenzie Delta is thought to have at least 9 Tcf of gas, and some at Canada's eternally optimistic National Energy Board suggest that there may be as much as 55 Tcf. Alaska is reported to have about 35 Tcf.[24] These deposits will

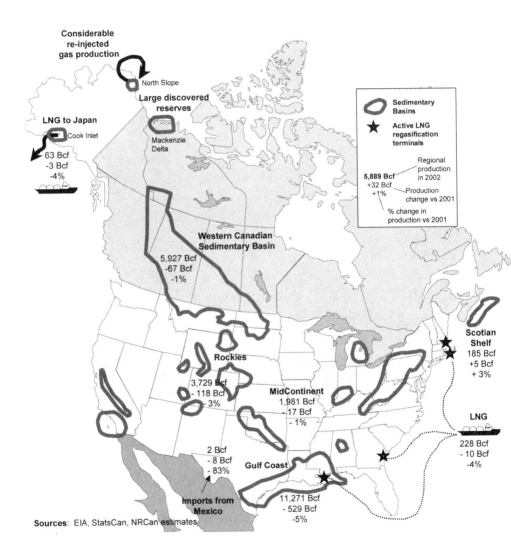

Figure 5.4. The distribution of major natural gas deposits (in sedimentary basins) in the United States and Canada. *Source: Canadian Natural Gas: Review of 2002 & Outlook to 2015, NRCan Canada.*

doubtless be helpful in the future but will not begin to make up for the kinds of falls now being seen, and there is no likelihood that they will be online before 2010, since they require the building of immensely expensive pipelines. The Canadian pipeline appears to be going ahead, although there may be delays owing to objections from the indigenous peoples on the pipeline route and court cases stemming from possible government irregularities.

By early 2004, the Alaskan pipeline was still mired in controversy, in part because the private sector wants a huge subsidy from American taxpayers, as well as price and market guarantees. There are immense pressures to solve this problem quickly, but it is part of the U.S. Energy Bill, which has been tied up in political knots for years. The last U.S. Energy Bill was passed in 1992, and critics argue it did little more than help the United States become further dependant on hydrocarbons. The next energy bill is designed to make the the country almost completely reliant on the rest of the planet within three decades, much as Japan is now. The *New York Times* called the oil and natural gas legislation "the most depressing" it had seen in recent history; others accused it of pandering to every polluter and offering pork to every lawmaker that might show some opposition.[25]

North Sea—United Kingdom vs. Norway

The U.K. part of the North Sea is a classic example of what happens when the desire for cheap energy, and later a deregulated, free market, meets a large gas play. Although the North Sea is most often connected with oil, in fact gas was found there (off the Yorkshire coast) in 1965,[26] well before the first oil fields in British and Norwegian waters in 1969.[27] By the mid-1970s, the southern North Sea was supplying most of Britain's natural gas. Later huge gas finds in the ferocious waters of the northern North Sea allowed Britain to become highly gas dependent and eventually to export to Europe by the late 1990s, which had, by contrast, fueled its central-heating revolution with oil.[28] This left Europe vulnerable to the oil shock of 1973, but Britain, once the dash for gas was fully underway, was much less affected. Things were rather different by 2003. Britain now has a chronic and serious gas addiction problem, while all of its sector of North Sea gas production is in decline, so much so that it will be importing Algerian gas, as LNG, in 2005.[29]

The contrast with the way Norway has managed its North Sea hydrocarbons is very marked. Norway has tried, with some success, to extract its oil and gas over a much longer period of time. The irony of the British free-market policy under Margaret Thatcher was that it prompted such enormous gas (and oil) production that it depressed prices and used up a huge resource very quickly. Now, as the world moves out of the petroleum century into an era of permanent energy shortages, Britain is turning to exactly the same gas exporters as everybody else, starting with Russia, North Africa, and the Persian Gulf.

The folly of this policy can hardly be exaggerated. In addition to producing greater gas dependence, the low prices also strongly discouraged the development both of alternative, renewable sources of energy and of conservation of the considerable resources Britain had discovered. In 2000, it was reported that there were only 2 megawatts of solar generation installed in the whole of the United Kingdom. Britain is not Nevada, to be sure, but even Norway, with less sun and far more fossil energy to spare per capita, has been way ahead of the United Kingdom in solar and renewable energy installations. Energy is at least partly a matter of attitude and will.

As U.K. North Sea gas production declines, Norway's gas production may increase somewhat and continue for many years, helping to supply a considerable part of Europe's lusty demand for gas. However, in 2001 for the first time, Norway produced more gas than it discovered, an ominous sign that unless it finds new gas perhaps in the Barents Sea, its reserves too will soon reach their limits.[30] In perhaps fifteen or so years, Norway is likely to run into the same trouble as Britain.

Indonesian Archipelago

Indonesia has been an oil producer for well over 100 years and was the target of the oil-starved Japanese in the Second World War. Ironically, it later became the largest of supplier of liquefied natural gas (LNG) to Japan and indeed in 2004 is still the largest LNG producer in the world. However, despite having large reserves reported at over 90 trillion cubic feet, Indonesia is experiencing production decline. This may be due to lack of infrastructure, especially in the case of the offshore Natuna field, which contains half its reserves. But Natuna is also a distant gas province and has a very high carbon dioxide content.

It may be that Indonesian production doesn't surpass the original peak, but if production remains at about the present level of nearly 2 trillion cubic feet per year, then Indonesia should still be producing considerable quantities of gas for at least another twenty or more years. However, the pressure to produce more gas by increasing infrastructure is strong, and Indonesia is very different from the more prudent Norway, not least in having fifty times the population. Looked at in world terms, no Southeast Asian gas producer has anywhere near the capacity of the "Axis of Gas,"[31] and almost all of Indonesia's gas is already being taken by Japan, South Korea, and Taiwan. As China becomes an enormous gas importer from the middle of the decade, only Malaysia and Australia have anything like the kinds of reserves to service such a giant, and even those will not be able to keep up for long. Energy is soon going to become as critical in Southeast Asia as it is North America.

Sunrise and Noonday Provinces

Russia

Now that the U.S. reserves have declined so much, Russia is reputed to have the largest natural gas reserves of any single nation,[32] but there are doubts about this. Its "public" reserves are, according the U.S. Department of Energy, 1,680 Tcf. It extracts just over 20 Tcf per year, and presently exports 6 Tcf and rising. Even this mighty production would not satisfy U.S. demand, which is over 23 Tcf per year, and also trying to rise. Russia, the great hope on which the planet is apparently relying for future gas, is stated to have more than double the reserves of Iran, which appears to have the second largest reserves. It must be a cause of concern then that Russian gas production is not expanding nearly as fast as oil production did. This is blamed on the company Gazprom, which controls the great majority of Russian gas reserves. It is claimed that its non-Western business and engineering practices mean that it does not invest in enough pipelines, infrastructure, and exploration.

However, a major reason why Gazprom does not have enough money to invest is that gas prices for Russian users are about one-tenth the prices in Europe, and are, according to the company, actually less than it costs to produce the gas. Furthermore, even at these very low prices, many customers either can't or don't pay their bills, to the tune of $2.7 billion in 2001 alone, Gazprom claims.[33]

Other factors that may restrict future gas expansion include cost overruns for ExxonMobil and Shell, which are developing huge projects on Sakhalin Island.[34] Located in the Russian far east just north of Japan, Sakhalin Island is one of the last, largely unexploited areas of oil and gas in Russia, and possibly the world. The high hopes for Sakhalin suffered setbacks at the end of 2003 when two of the blocks (4 and 6) under exploration were abandoned by BP and Rosneft after failing to find major oil and gas reserves.[35] At the beginning of 2004, the Russian government failed to give a license to ExxonMobil and Chevron Texaco for development of the Sakhalin-3 project, which they had been negotiating since 1993.[36] Overshadowing all oil and gas production in Russia is the case of Yukos, and the arrest of its chief, Mikhail Khodorkovsky, which is causing foreign petroleum super-majors to think carefully about their investment strategies in Russia.[37] It should also not be underestimated that both local and international environmental groups are determined to stop the major oil and gas developments by Exxon, Shell, and others on Sakhalin.[38]

Finally, though no doubt Gazprom could be run better in many ways, there are also some concerns that Russia's gas reserves, though very large, may be considerably smaller than quoted. Russian gas production relies very heavily on western Siberia, particularly its largest gas field, Urengoi. This field is now estimated at 230 Tcf, down from earlier estimates of 350 Tcf, and at least half its gas has been extracted. Furthermore, some analysts believe that despite a recent large, new development,[39] western Siberia faces serious gas-depletion problems and will start to contract by more than 5 percent per year as soon as 2003 or 2004.[40] Analysts Wood Mackenzie estimate Russian reserves at 1,316 Tcf, which is more than 350 Tcf less than the EIA estimate. Furthermore, even this may be high, since (as explained in chapter 2) Russian reserves classifications are about 30 percent higher than their Western equivalent.

In a world hungry for massive and growing quantities of gas, what matters is not just relative size (though this is politically very important), but also absolute capacity. The reduction in Russia's estimated gas reserves are therefore serious. Location is also an important matter, and in this sense, all the largest gas deposits have the unfortunate characteristic of being a long way from where the gas is wanted. In many cases, these nations are run by administrations that Western governments may see as threatening.

South America

Because the United States is the dominant user and importer of gas in both North and South America and is also facing the greatest problem with gas supply not meeting demand, the discussion of Central and

South America will be from a U.S. perspective.[41] There are lessons here for all governments and citizens everywhere in the world, however, and the case of Bolivian resistance to exporting gas to the United States will be looked at in more detail, though the account will necessarily be incomplete since matters are still unfolding.

South and Central America have long been of the greatest strategic interest to the United States because of the presence of oil. Mexico has been pumping oil to America since the First World War, as has Venezuela. Between them, they pump almost as much oil to the United States as Saudi Arabia and Canada combined.[42] In addition to oil, Mexico was for some years a natural gas exporter to its northern neighbor. Since the late 1980s, however, Mexico has become a gas importer, so that by 2001 the United States exported about as much gas to Mexico as it imported via LNG.[43] It seems unlikely that this situation will continue for long, since the United States will be growing increasingly anxious for all the natural gas it can lay its hands on.

In fact, the situation is much more complicated than a first glance would suggest. Mexico may yet have quite a lot more gas to extract,[44] but its nationalized petroleum company, Pemex, is not regarded as one of the most "efficient" at extracting resources as quickly as its northern neighbor and NAFTA partner might wish. This may be a good thing for Mexico in the longer run, since it will leave more in the ground for times when desperation is the order of the day. Mexico also wastes a lot of its gas by simply flaring it at the well site. In September 2003, the U.S. Department of Energy predicted that Mexico would start exporting gas to the United States in 2019,[45] a rather extraordinary statement by any standards, especially as oil and natural gas predictions in these almost certainly permanent "tight market conditions" are now little better than long-range weather forecasts. As a sign of the volatility of the situation now, just three months later, in December 2003, the DOE predicted that Mexico would not become an exporter at all, but would continue to be an importer till at least 2025.

In the meantime, Mexico's natural gas hunger is only set to grow, expecting to double its 2000 demand by 2010, so that it will be using almost one-fifth of what the United States is using now—a very large amount of natural gas. This demand expansion, which is likely to lead to unfortunate political and strategic developments, is occurring for two different reasons.

First, gas demand is increasing directly as a result of a growing population[46] that is trying to "develop" economically, and second, indirectly, because Mexico is set to become a gas-transit country, taking LNG for the United States and piping it north either as gas or as electricity. If the latter, then so much the better for the United States, because Mexico gets saddled not only with unpopular and potentially explosive LNG terminals, but also, if the electricity is generated south of the border, Mexico has to deal with the associated pollution and greenhouse gas emissions. This may be quite a problem, since Mexico has signed the Kyoto Protocol, unlike the United States, which famously withdrew from the process in March 2001.

Traveling south, Venezuela has large gas reserves,[47] second only to the United States in the Americas. Colombia has only relatively small deposits.[48] As of 2003, Colombia, previously a modest user of its own domestic gas production, was getting ready to increase its natural-gas consumption significantly.[49] The plans of 2002 called for building a natural-gas pipeline from Colombia to Venezuela. But, curiously, in its early stages it would export gas from the minnow Colombia to the giant Venezuela. However, the ultimate intention is to send gas the other way. Venezuela, with stated reserves of 148 Tcf is only slightly smaller than the U.S. reserves. Even so, it extracts gas at barely more than one-twentieth of the U.S. rate. It can be no surprise that Venezuela has discussed building LNG terminals with commercial LNG companies such as Shell and Mitsubishi. Ironically, these discussions have slowed because of the political problems caused by the very nation that would get the most benefit from the LNG. The real irony, however, is that the plan to remove President Hugo Chavez and install a highly pro-U.S. president backfired. Chavez had not even shown unwillingness to sell to the United States, but his anti-free-market political ideology was not to Washington's taste. Washington may have to be willing to swallow its dislike, because the U.S. consumer will soon be asking where the gas is, and Venezuela still has quite a lot of it.

Venezuela may face another problem, which other producers face too, and that is if they extract too much associated gas (gas found with oil), too early, they may harm the prospects for their oil production. This can happen, because the gas cap so often found above oil deposits helps pressurize the reservoir and increases the oil flow. It usually only makes sense to extract more associated gas when the oil output is well into decline, and of less interest. As gas becomes more and more valuable in North America, the political and commercial pressure to extract associated gas sooner is likely to increase, making the task of the petroleum reservoir engineer that much harder.[50]

Venezuela, then, remains the most obvious gas exporter in terms of size of reserves, but a rather problematic one, in the short term at least.

The islands of Trinidad and Tobago may be geographically small, but since their first shipment in 1999, they have grown rapidly to become by far the largest exporter of LNG to the United States.[51] In 2003 they were exporting 29 Bcf per month on average (though there are considerable variations month by month).[52] New LNG trains to ship ever increasing amounts of gas are being built, but this has given rise to worries that Trinidad's gas reserves[53] may be overstretched by such prodigious demands, and that it is committing too much of its gas to LNG. In 2002, 36 percent of its gas was exported, but there are estimates that when the fourth LNG train comes in stream 2006 (and two more trains are being considered), it may be using as much as 80 percent of Trinidad's gas production.[54] Even the lowest estimate is more than 50 percent of production. Trinidad's domestic economy is heavily dependent on its natural gas. Ninety percent of all the energy that

Trinidad uses comes from natural gas, and a great deal of gas is used to make methanol and ammonia, especially for nitrogen fertilizer. Trinidad is the world's largest exporter of both methanol and ammonia.[55] Trinidad's remaining gas reserves are thought by some to be about 30 Tcf,[56] and recent exploration drilling has not been promising.[57] At the kinds of increasing extraction rates now happening, Trinidad's gas will not last much beyond twenty years. After that, it will be in catastrophic energy and economic difficulties, and with a fast-rising population to fuel and feed.

There is another strange complication with gas in this region, which is a direct result of the huge long-term investment in LNG export facilities in Trinidad, without enough gas to keep it supplied beyond two decades. In fact, Trinidad shares a gas province with Venezuela, the Deltana, which some have suggested will be needed in future to keep the LNG trains supplied. This is not likely to be favored by Venezuelans, who are considering exporting the gas themselves. It has been conjectured that this matter is of such importance to the U.S. gas market, that it is one of the reasons for past and continuing efforts to remove President Hugo Chavez.[58]

The rest of South America is, like Colombia, planning to increase its use of natural gas—in many cases by a great magnitude.

Chile is perhaps the most aggressive, and possibly most foolish, case in point. It increased its gas use by almost a factor of five from 1991 to 2001, and plans to increase by another factor of four by 2011, yet it has reserves of only 3.5 Tcf, and falling production.[59] It is now totally dependent on Argentina for gas imports, which amounts to over 80 percent of Chile's gas consumption. This has made it vulnerable not only to Argentinean currency problems, but also to other man-made and natural problems, such as labor unrest and the landslide that caused power outages in Chile and Argentina in 2002. It has also found now-familiar problems with overly optimistic ordering of natural-gas electricity-generating power plants, which are then canceled, leaving the pipeline operators with a bad investment. This suite of problems high-lights the underlying and disagreeable fact that everybody wants to con-sume more gas, but few can produce enough to do so.

In South America, Bolivia has the largest amount of gas to spare: it is estimated to have somewhere between 24 Tcf and 52 Tcf of gas reserves,[60] and everyone has their eyes on it. In fact, Bolivia has much more gas than Mexico, at least according to current estimations it is 8.8 Tcf (which may be too miserly).[61] Peru also has generally untouched reserves,[62] and Argentina has sizeable reserves,[63] though both have con-siderably larger demand than Bolivia, and plan to increase their reliance on gas considerably.

The giant in the region is Brazil, which is desperate to become energy independent and has tried for a number of years to develop a stable relationship with Bolivia for importing gas.[64] Various problems have continually upset most plans, ranging from currency difficulties, more overordering of gas power plants (in reaction to a drought-induced power crisis in 2001) and then cancellations, and the extraor-dinary complexities of pipelines and transportation. There is also the matter of desecration of a whole section of what remains of the Amazon. However, Brazil's gas situation has recently changed dramat-ically, with the discovery by its own petroleum company, Petrobras, of large offshore gas deposit in the Santos basin estimated to contain 14.8 Tcf.[65] Petrobras feels certain that this is just the beginning of many more finds, though others are more cautious. Nonetheless, this find almost doubles the EIA estimate of 8.1 Tcf for Brazil's reserves. However, Brazil's gas consumption is rising fast,[66] and unless it either caps its gas-demand growth or discovers still more gas, it will find itself, in some years hence, once again looking for gas imports.

In general, the nations of South and Central America are digging themselves into a carbon hole from which they will be hard put to extricate themselves. Their case, of some countries with quite large gas reserves but with apparently uncontrollable appetites and many social, political, economic, and environmental factors wreaking havoc with their overambitious and probably unwise plans, should serve as yet another warning to the world. Even more profoundly, the case of Bolivia provides further lessons for citizen and CEO alike.

In 2002 and 2003, Bolivia was under severe pressure from Washington and the IMF. Its then president, Sanchez Gonzalez de Lozada,[67] also known as "El Gringo," because of his U.S. ties, was trying to force through a pipeline to connect the large gas fields of Margarita in the south of the country to the Chilean port of Arica, where it would be liquefied and sent mainly to the United States. The War of the Pacific,[68] which cut the Bolivians off from the ocean in 1883, has left them with an abiding hatred of the Chileans. This is now compounded by the American so-called war on drugs, code-named Plan Dignity in Bolivia, which has seen brutal armed forces eradicate 80 percent of Bolivian indigenous peasants' only real source of livelihood, their coca fields. The promises of aid to replace this loss were, as they so often are, false. But the Bolivian native people, so badly treated by at least five hundred years of colonial invasions and conquests, rose up in the late summer of 2003 and forced Lozada from power, chiefly because of his efforts to export gas.

The importance of the Bolivian story, however it plays out, can be stated simply: Those that have natural gas had better beware. Those that don't will need to seek out fruitful relationships and accommodations with those that do. The other course for gas-poor countries is not to become gas-addicted, but for many countries it's already too late.

Axis of Gas

There are many reasons for suggesting, even begging, that the countries of planet Earth reconsider expanding their natural gas consumption. Further reasons will be discussed in chapter 7, but here in the supply chapter the cases of the other great gas provinces of the world will show that while there is gas to be had for the moment, this period will be short-lived. Most of the countries that have large remaining gas reserves are in politically sensitive or dangerous conditions.

Having now considered some of the possibilities and difficulties of natural gas supply south of the U.S. border, it remains to look at why the natural gas supply may never be really stable or reliable for much of the world. Those with greatest market power will be better served, as usual, but even they will be in difficulties relatively soon.

The reason for what some may regard as a pessimistic view of gas supply stability is that the overconsuming, highly industrialized world, with the exception of Australia and Norway, has used most of its cheap and easy gas, just as it has done with oil. This now leaves Russia, the Middle East, the Caspian basin, and Africa to fill the supply chain. All of these regions face formidable political, economic, and environmental problems, and most of them have population problems of fearful proportions as well. Their population problems are second only to those of two of their largest potential new customers—India and, especially, China.

The Caspian Basin

The Caspian Sea region includes parts of Russia, Kazakhstan, Turkmenistan, Iran, Azerbaijan, and, though it is not on the sea itself, Uzbekistan.[69] Currently the region is a relatively minor producer for political, economic and infrastructural reasons, but the EIA estimates that the Caspian region has up to 4 percent of the world's total natural gas reserves with 232 Tcf of proven reserves. Others however, offer considerably lower numbers, with a total of 174 Tcf.[70] In 2001, Uzbekistan produced 2.23 Tcf, which was almost as much as the other Caspian nations combined.[71]

The Caspian has been a region of intense political conflict between great powers for at least two centuries—long before oil was such a vital commodity. In the early 1990s Caspian oil reserves were thought to be as large as those of Saudi Arabia, perhaps even rivaling those of the Persian Gulf. Though very high figures remain in the EIA's outer estimates, few now believe that the Caspian basin will live up to its promise, or at least the promises that were made on its behalf. Thanks in part to inflated estimates for its oil reserves, it has seen an extraordinary military buildup of foreign forces since the fall of the Soviet Union, particularly those of the United States.

The infrastructural limits on Caspian gas production are, as so often, mainly to do with transport, or lack of it. There are limited pipelines in this area. Most are owned by Gazprom and go out through Russia. Though Asian demand seems likely to outstrip that of Europe and thus favor eastern exits, the United States would like to see the gas come out in Turkey so as to bypass the Russians. This, along with desires for oil, may have precipitated many international events in the past few years including the war against the Taliban in Afghanistan and the Georgian velvet revolution, among others.

In 1997, Turkmenistan and Iran completed a pipeline[72] linking the two countries, thereby becoming the first (and so far, only) natural gas export pipeline from Central Asia to bypass Russia. There is a proposed natural gas pipeline called the South Caucasus natural gas pipeline (the Baku-Tiblisi-Erzurum or BTE). There are also proposals for a trans-Afghani pipeline to get the gas to markets in Pakistan and India.

In the Caspian region, it is very hard to predict the future of gas supplies, or indeed almost anything else.

The Persian Gulf

As with oil, the Persian Gulf is reported to contain more natural gas than any other province. It is slightly more than that of the former Soviet Union (FSU),[73] but unlike the FSU, whose reserves are spread out over a much larger region, the Gulf reserves are concentrated in a relatively small area, covering only 7 percent of the surface of the Middle East, which Simmons calls the Golden Triangle.[74] The Persian Gulf has been wracked by wars, invasions, and occupations for decades. It is hard to imagine that any of the colonial and empire powers, such as Britain or the United States, would have shown much interest in, or spent billions of military dollars on this region, had it not been its dreadful misfortune to contain the world's largest deposits of the most vital commodity—highly dense, stored energy.

Where Saudi Arabia is the Gulf giant of oil, for gas that title is shared by Iran and Qatar.

Iran

Estimates of Iran's natural gas deposits vary considerably from 812 to 940 Tcf.[75] Iran has part of the largest single nonassociated (no oil) field in the world, at South Pars, and this reservoir alone is thought to contain somewhere between 280 and 500 Tcf. If the higher figure turns out to be correct, then it will contain more than half of Iran's entire reserve. Little of South Pars has been touched. Iran plans to spend billions of dollars to increase the amount of gas produced and might like to export it, especially as Iran's conventional crude oil output is falling (mainly for geological reasons), and its other main exports of carpets and pistachio nuts are not quite so vital for world industrial growth.

Natural gas supplies half of all Iran's energy needs. In 2001 Iran produced about 2.2 Tcf of gas, most of which was used by its own population. The most significant problem for natural gas in Iran, apart from the obvious difficulty of getting the gas to market, is that much of Iranian gas is sour, meaning it contains more than 1 percent hydrogen sulfide, which, being poisonous and corrosive, must be removed before the gas is put into pipelines and delivered to the user. Iran is seeking to export its gas both by pipeline and by LNG, but apart from a pipeline to Turkey, most of the necessary transport infrastructure is still in the planning stage.[76] Although Iran may appear potential to become a large gas exporter, there are four important factors which should be taken into account.

Firstly, despite its large reserves, Iran is a actually a net importer of gas at the moment, because so much of its own gas is far from the Iranian people who would like to use it. Thus, it receives gas from its neighbor Turkmenistan. To complicate matters, Iran is in dispute with all the nations that surround the Caspian Sea,[77] over what percentage of the Sea it has a right to exploit. Turkmenistan is one of those nations, and it also shares a long border with Iran. As indication of how serious these Caspian disputes can become, the EIA notes that "On July 23, 2001, tensions flared in the Caspian Sea region when an Iranian gunboat intercepted two BP oil exploration vessels off Azerbaijan's coast. Following the incident, BP suspended exploration in the disputed block."[78]

Secondly, and in the long run most likely of greater significance, Iran's population is growing fast, and helping to cause the domestic economy to demand ever greater quantities of gas. To avoid destabilization, the Iranian government will have to consider carefully how their gas is used. They may, for instance, have noticed what happened in Bolivia in 2003.[79]

Thirdly, Iran may decide to take a large proportion of gas from its South Pars field and re-inject it into its ageing oil fields, in order to try to bolster its oil production. With oil at over $40 a barrel, economic factors for this course are almost overwhelming.

Fourthly, there are doubts being raised as to whether the South Pars (and contiguous North Field) really is so large.

Qatar

Although a tiny country, Qatar's gas reserves are enormous. Estimates vary wildly from a "low" of 509 Tcf to a high of 910 Tcf.[80] Most of Qatar's gas is in the North Field-South Pars offshore reservoir, which it shares with Iran. Qatar is the most gas dense country in the world and, appears to have the third largest gas reserves after Russia and Iran. The reserves are more than two and a half times greater than the remaining U.S. reserves. It is already exporting LNG and plans to increase its output greatly. By no coincidence, Qatar's capital, Doha, sustains a heavy U.S. military presence and was vital in the invasion of Iraq. The same caveat regarding reserve over-estimation applies to Qatar.

Saudi Arabia

While having the world's largest oil reserves, Saudi Arabia has much less gas—at least by comparison with some of its neighbors. Nevertheless, in absolute terms, its gas reserves are enormous and rank it in the top ten. None of Saudi Arabia's gas is currently being exported, and to date it has no plans for developing LNG.[81] However, in December 2003—at a U.S. LNG summit, recognizing that a global shift to gas was taking place, Saudi Arabia's minister of petroleum said that he "would evaluate all options—including LNG exports, in order to maximize the benefits to Saudi Arabia's economy."[82]

Kuwait

At 52 Tcf,[83] Kuwait has enviable reserves of gas, though small by comparison with most of its neighbors. Despite these significant reserves, which it does intend to develop, Kuwait has been a substantial importer of gas in the past, particularly from Iraq. It is actually planning to increase gas imports.[84] Kuwait uses gas for electricity production, and also to avoid burning diesel, which allows it to increase oil exports.

Iraq

Iraq is reported to have over 100 Tcf of gas[85] but produces very little of it now, though it once reached an output of 700 Bcf in 1979. Under

Saddam Hussein it had intended to build an LNG terminal for export purposes. However, there is official and unofficial talk of major damage to Iraq's oil reservoirs,[86] possibly as a result of trying to extract gas rather than oil. How this will affect future production (of either gas or oil) is hard to know. With Iraq under foreign occupation, it is very hard to predict what the future holds for any aspect of the country. Late in 2003, guerilla resistance to the U.S. forces was stiffening, and the situation seemed to be descending into a scenario reminiscent of Vietnam.

Africa

From the instabilities of the Middle East, attention shifts to a continent that historically has suffered badly from the attentions of European colonizers. As Libyan oil has long been in decline, Africa's main oil reserves are now concentrated in the Gulf of Guinea,[87] with Nigeria as the largest, followed by Angola. However, Africa's natural gas reserves are much more plentiful than oil, and becoming much more interesting to a gas-hungry world. Unfortunately for those wishing to extract them at speed and in quantity, rather than being concentrated around a bountiful basin such as the Persian Gulf, they are spread out across many of the northern and western coastal countries of the continent, from Egypt through Algeria all the way around and down to Angola. Africa will therefore be a continent of expedience, of interest in the first decades of the millennium, not because it is more politically stable than the Caspian or Persian Gulf, but because it is much weaker and more malleable than the recalcitrant states to the northeast in the Persian Gulf, and also because it is physically closer to the United States.

With the exception of the gas giant Algeria,[88] which was the first country in the world to export gas via LNG tankers,[89] most African countries have barely begun to produce gas for the market. Nigeria, with the second largest reserves, has flared most of the gas so far extracted along with its oil. Increasingly, Nigeria will capture the wasted gas and export it, most likely to the United States, via LNG tankers.

The other two main countries with large and mainly untapped gas reserves are Libya[90] and Egypt,[91] with Egypt now ahead of Libya in proven reserves. How the presence of a hydrocarbon bonanza next to Israel will affect the politics of the region remains to be seen. Such potential wealth always has an effect, and it is rarely beneficent, especially for those poor countries that came to global capitalism rather late and by coercion.

There is little doubt that Africa will become a major gas exporter. The top four gas states, mentioned above, all have more, or indeed far more, gas than the North Sea, which was itself a very large find of natural gas (and oil). It is safe to say that Africa will be exporting large quantities of gas in the early decades of the twenty-first century. Whether it will bring any of those African states any good remains to be seen—unless they defy tradition, most of the exporting countries will use the ensuing wealth to enrich their rulers and further dispossess and enslave the poor. Such is the curse of the liquid hydrocarbon. Oil has been called the devil's tears. Natural gas is perhaps the devil's breath.

Southern Asia

This still leaves the fast-growing economies of southern Asia, which include the most populous nations in the world, China and India. Unfortunately, Indonesia, the largest gas producer among their number, had entered production decline, at least it appeared so in 2003.[92] The Asian Tigers, though recently tamer than they once were, must turn to Malaysia and Australia as the most serious hopes for rising supplies of local gas, and these two countries are indeed increasing their gas production as quickly as they can. Australia has signed a long-term contract with China and is being courted by many others. It plans to become a world-class LNG exporter, and for a while, it will be able to achieve that. But the total gas reserves of Indonesia, Malaysia, and Australia are barely half of tiny Qatar's reported endowment. There will come a day when southern Asia will be left gasping for gas, and they will have to get it from the Axis of Gas.

The outside hope is that the South China Sea proves to be an oil and gas bonanza like the North Sea, or bigger. Years of contention over access has meant that little drilling has been done to confirm this one way or the other. Some petroleum geologists are skeptical as to whether the source rocks that may be present have actually been through the oil and gas window, which would mean, in plain terms, that there would be no oil or gas to extract. The fact that China has laid claim to all the Spratly Islands, while six other nations are interested in all or some of them, suggests both that desperation has set in and that many are interested in seeing whether this is the last Aladdin's energy cave recoverable with known technology.

For the moment, China, the rising powerhouse of the East, is just maintaining gas production ahead of consumption, but this is set to change in 2005, when China expects to become a net gas importer. It started importing oil ten years ago, in 1993, and already imports over 30 percent of its oil, passing Japan in 2003 as the world's second largest consumer of oil, and expecting to consume as much oil in 2010 as the United States does now. This is leading to increasing alarm in security and strategic circles that one day China will be challenging the United States for Earth's dwindling oil supplies. By 2010, if China has its way, it will also be a massive importer of gas, since it intends to rise from using 1 Tcf a year (in 2003) to anywhere between 2 and 4 Tcf, partly in answer to critics of its coal-burning practices, which supply so much of its energy. This will also put immense pressure on the world gas supply system, since so many other countries will be vying for the same gas.

Looking geographically at how the world's gas supply will develop, if any continent is to be considered a sunrise producer of natural gas, it would be Africa. Looking west from the Axis of Gas reveals the great sunset producers of North America and the North Sea; east of the Axis of Gas, the declining gas giant Indonesia watches the rise of Australia, whose gas reserves are almost as large as Indonesia's. Officially Iran and Qatar may have nearly ten times as much gas as either Indonesia or Australia. The use of African and Australian gas will be strategically useful to those charged with energy procurement policy. For the world, however, the existence of new and fairly compliant gas territories will help condemn us to a false path of greater hydrocarbon addiction and will involve the investment of trillions of dollars in pipeline and LNG infrastructure, when that money should have been spent on moving the world to renewable energy and reducing global aggregate demand rather than encouraging it. As a final grim epitaph to human energy mismanagement, if the doubts about Russian, Iranian, and Qatar's reserves are shown, by the drill bit, to be well founded, then building a global LNG network will turn out to be the greatest single economic folly in history.

6

Where on Earth Are We Now?

The Gas Showdown Begins

In 2003 the U.S. Secretary of Energy announced that the United States was globalizing the gas market. Put crudely, this means that the United States is about to connect itself permanently and massively into a global natural gas production and distribution system, which doesn't yet exist. This has immense implications for the world. By 2003 it was clear that the United States, Canada, and Mexico were in severe difficulties with gas supply, and that the United States and Canada had most certainly passed their production peaks and were both in decline. Depletion will thus attend their natural gas policies forevermore.

What is the world situation and what are governments doing about it, if anything? What are the geopolitical ramifications, the role of prices, free markets, and deregulation in exacerbating the problem?

It is impossible to predict exactly when global natural-gas production peak will occur. Much of the infrastructure to bring east Siberian gas to users has not been built, nor have many of the pipelines from future or rising gas-producing regions, such as Alaska. Although the major petroleum corporations have declared their intention to turn to natural gas as their primary fossil fuel of the early twenty-first century, no one can say how quickly oil will be replaced by gas, and to what extent, given the enormous infrastructure changes that will be required. If there is no economic slowdown and a global dash for gas occurs, and if the current conventional reserves are not greatly augmented, or have not been overestimated, then global gas peak could occur by 2020. On the other hand, if any of these factors change, and it is quite possible that they might, then global gas peak may be much farther off—or much sooner. In any case, it will resemble a plateau, rather than a sharp peak.

To some, this might look like a formula for carrying on with the changeover to natural gas, and reassessing the situation in a decade to see how things are going. There seems almost no question that this is what many governments will do, especially (and ironically) those most committed to the Kyoto Protocol. The many reasons for not changing over to gas, and attempting to use much less energy in society as a whole, are definitely not going to appeal to orthodox policy makers, economists, and ordinary Western consumers.

In a sense, most of the rest of this book is for the "walking worried," those who realize that if, once again, we wait until the last minute, it will be literally and physically impossible to build a low-energy infrastructure to replace the high-energy one we now have. As a final attempt to persuade the orthodox, the example of the United States is the most compelling. As a place that, with a few exceptions, refuses to reduce its energy demand, the United States is now in a position somewhat analogous to Sparta some 2,500 years ago, in which it had to devote all its resources to hold on to the nation it had enslaved. Energy functions as a kind of slave, and the United States is increasingly at war with the world, primarily because of oil, but soon that will include natural gas. All nations that find themselves in this position eventually crack under the load. The United States may be much closer to cracking than most realize, but then again, as long as the military option keeps working and there is another energy-rich country to attack, the situation can lurch on a bit longer. Given that the majority seems to act only when the avalanche is upon the roof, it is quite likely that no prediction, however accurate it is, will be sufficient to shift mainstream policy making or opinion. Thus, it is only those who think that we have already gone too far who will be willing to act, make the kinds of big changes required, and more than anything start building a new infrastructure while we can.

Another major reason why we shall wait until it is too late to change our energy path is that prices are not telling or reflecting the coming shortages. Even if economists had the methods to reflect more of the worth, scarcity, and problems of hydrocarbons, their position as the very lifeblood of industrialism means that they are far too political to allow prices to have free rein. For instance, the price of oil, and soon gas, reflects none of the military costs involved in securing the supply. But beyond the military reasons that energy prices fail to tell the truth, there is also a structural reason.

Prices, like economists, know nothing about geology or Hubbert curves. Even if it can be accurately predicted that oil or gas supply will become constrained, unregulated price can only reflect what the "market" is willing to pay. Indeed, as a substance gets rare, very near to extinction or exhaustion, its value may very well go up, increasing the likelihood that it will be even more sought after by producers, thus hastening its decline. Since economics assumes that everything is either produced or grown in some way, or else can be substituted, it can and generally will drive a limited supply to destruction. However, because of the versatility and energy density of liquid hydrocarbons, humans will be forced to confront the decline first of oil, then of natural gas. This confrontation will come first in North America, where thanks to vast and unquenchable demand, the United States, followed by Canada and Mexico, will enter the two states of oil and gas decline at roughly the same time.

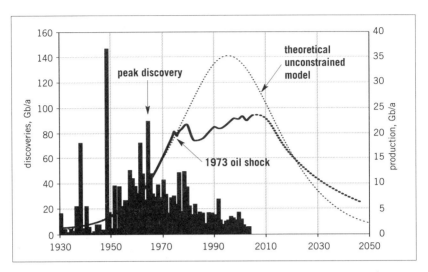

Figure 6.1. World conventional oil discovery, with actual and unconstrained production, past, present, and future. The black bars show discovery; the dotted line shows how production would have behaved without the 1970s oil shocks; the solid line shows actual production, and prediction. *Source: Colin Campbell.*

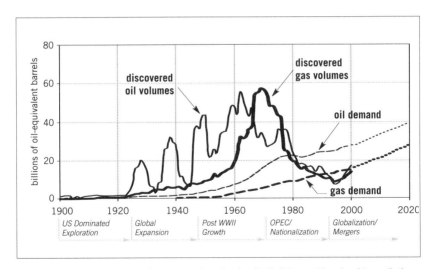

Figure 6.2. World oil and gas discovery and production (called demand here), with predictions, according to ExxonMobil. The only real difference between Exxon and the "peak" petroleum geologists, at least one of whom used to work for Exxon, is that oil demand cannot continue as projected. Projected gas demand/production is too optimistic. It is very hard to predict how much further and for how long gas production will rise. *Source: ExxonMobil.*

What Is the Relation of Gas to Oil Peak?

The phenomenon of global oil peak gained currency in the first years of the new millennium. The pattern of global gas discovery, examined in chapter 5, looks remarkably similar to that for oil. As figure 6.1 shows, for oil there was a peak of discovery in the early to mid-1960s followed about forty years later by an inevitable peak-plateau of production. There is still debate about whether global oil peak is happening or about to, but not many petroleum geologists put conventional oil peak beyond 2010.[1] Many think that we are now already seeing the effects of oil peak, and indeed the major oil and gas companies are announcing that they are increasingly switching to gas as their primary fuel of production. Shell has even predicted that gas demand will outstrip oil by 2025.[2] In fact, the graphs of oil and gas discovery suggest that gas will overtake oil before then, but driven by oil supply difficulties rather than "pure" market demand. If gas extraction peak were to follow oil peak at the same distance of about forty years after discovery peak, that would make 2010 the peak. This conceivably could happen. Most of the easy gas has been extracted from the planet, and all future gas is a long way from where it will be sold. This will hamper production very considerably, making it much slower and more expensive than oil. Thus gas extraction will plateau, but its timing may depend most on whether reserves have been over-estimated.

The pictures of discovery, especially the one offered by ExxonMobil in figure 6.2, should deliver a sobering dose of reality to those with cornucopian beliefs. If the world's largest nonstate oil company, known for being highly aggressive in business, antienvironmental in its activities though not in its public relations, and willing to exact high human cost in its quest for more oil will show numbers that effectively declare oil to be a sunset industry, then those who believe we shall find lots more oil and gas, or some wonderful substitute, will soon be forced to admit that the industrial economy will not run on promises, even if the futures market appears to.

Oil to Gas Price Relationship

The prices of oil and natural gas are linked, though far from locked. A barrel of oil typically contains about the same amount of energy as 5,500 cubic feet of gas. The wellhead price, at least in a mature market such as the United States or Britain, usually reflects this, so that oil at $30 a barrel sees gas around $5 to $6 dollars per thousand cubic feet. The relationship is quite elastic, because many users cannot simply switch from one to the other. The main users who can switch are the big consumers such as power stations, which are often dual-fueled with oil or gas, though many of the new gas-fired power stations are not able to switch, an "economy" that by 2003 had already been shown to be shortsighted and expensive.

Energy and Empires

Since the rise of hydrocarbons as the engine of industrial growth, energy production has become a massive and centralized enterprise, requiring enormous infrastructure, investments, and bureaucracies. Although the East India Company and Hudson's Bay Corporation were very large, and at times effectively controlled whole countries, the first global corporation as we know them now was arguably Rockefeller's Standard Oil. It was eventually broken up because of its monopoly,[3] but some of the dismembered parts have been rejoined by corporate surgeons, especially Exxon and Mobil, which, with well over 50 percent of the original business between them, were the two largest parts of the original Standard Oil.[4] ExxonMobil is still the world's largest nonstate-owned oil corporation.

Standard Oil embodies the essential paradox of big energy: it has to build a huge infrastructure with many fixed costs, and therefore must find some way of guaranteeing a market and prices; otherwise it will not see sufficient return on that investment to satisfy shareholders. The only way to ensure that situation is to have a monopoly. Since private monopolies are obviously unfair, and supposedly prohibited even in such a bastion of capitalism as the United States, such a monopoly must be created covertly. Rockefeller and Standard Oil created that monopoly by any and every means that they could get away with. Despite the breakup of Standard Oil in 1911, nothing essential has changed in the way global corporations operate, especially the giant energy corporations. Energy companies from Standard to Enron are essentially operating the same game. Occasionally they get caught, either because they become too flagrant in their abuses, or else because the international stock market bubbles on which they depend eventually deflate. The great bull bubbles of the last 400 years[5] are reaching their climax, and when the expanding energy supply on which they now depend becomes constrained, then there are real reasons to project that the Industrial Revolution will begin to unwind. How this will play out, it is hard to say, but most historical examples of empires collapsing are not encouraging.

Understanding the nature of big energy monopolies is important because it is highly likely that quite soon big energy will not be quite so big, and those who have delayed building their local energy infrastructures will find themselves in the company of the "foolish maidens" who, rather appropriately, burned all their oil before midnight.

Two brief but highly significant examples are offered here of how "free" markets and monopolies conspire to addict the public to an abundance of cheap energy or promise them even cheaper energy, then either gouge them when the time is ripe, or leave them at the mercy of foreign energy suppliers when the local supply runs out.

A Nasty Lesson from Britain

The story of British North Sea gas was alluded to in chapter 5, with some explanation of how it came about that enormous gas (and oil) reserves were squandered, and at low prices created in part by overproduction. The low prices and the deregulated, free-market structure introduced by Margaret Thatcher ensured that little of the benefits went to the British people, who will suddenly find themselves facing natural gas imports beginning in 2005. To compound the matter, Britain is set to begin importing oil in 2007. Yet in 1999 the United Kingdom was the eighth largest producer of oil in the world,[6] and among the top ten exporters. It was also the fourth largest gas producer. This shows how quickly the tide can turn when the peak is reached, as was the case with both oil and gas production in the United Kingdom around 1999. The trade papers were full of reports that the United Kingdom had never produced so much oil and gas before. This of course is exactly the point about peak production. The mistake is to make an everlasting linear extrapolation from the rise of production and assume that it will go on rising forever.

Since industrial systems require big centralized, capital-intensive energy, monopolies will always be required to guarantee prices and markets, as previously explained. Thatcher handed the oil and gas companies the North Sea on a plate, whereas the Norwegians have kept state control over their petroleum resources. Monopolies are always dangerous entities, but far from deregulating them as Thatcher did, where they must exist, it would be wise to insist on some form of democratic control, as in Norway's case.

Britain converted en masse to natural gas in the 1970s after the discovery of gas first in the southern North Sea, then in the northern North Sea. This is often referred to with the typical Anglo-Saxon penchant for rhyme as the "dash for gas." Unfortunately for Britain, its reserves, though large, turned out to be limited by nature, and it will be forevermore a gas importer after 2005. For Britain it will be the second of three nails in the coffin of energy independence. The first nail was closing much of the coal industry during the 1980s and 1990s, partly as Margaret Thatcher took revenge on the miners and other unions,[7] but also because of reserve depletion and attempts to reduce greenhouse gas and other emissions. In an ideal world, no coal would be used, since it kills and ruins all the way along its production path. However, far from being reduced, coal use is increasing around the world. As the new millennium advances, some British voices, even those in authority, are starting to realize and say that Britain, and indeed all the British Isles, including Ireland, are going to face very severe energy problems before the end of even the first decade of the twenty-first century. This realization will likely be as momentous in its way as the fall of Communism was in Russia and its satellites. The rest of the world will do well to watch Britain to see how it deals with the kinds of changes that every nation and place will face during the first decades of this century.

California in the late 1990s

In California, the state government, in collusion with private energy companies, enacted widespread deregulation, removing many vital rules that constrained commercial behavior in the public interest. With deregulation in place, companies such as Enron forced up prices for electricity and sucked billions of dollars out of California's economy, until the bubble burst in 2001. With the installation of Arnold Schwarzenegger as governor in 2003, backed by the same people who brought in the original deregulation system, it looks highly likely not only that Californians will recover little if any of the money they lost in the 2001 energy crisis but will see even more deregulation. With an ever-tightening natural gas supply and increasing reliance on gas-fired power stations, they must be prepared for a repeat of 2001, with even more blackouts and potential damage.

The cases of Britain and California starkly highlight the same basic and inescapable problem of the need for monopolies when dealing with big energy[8] and the abuse that ensues when the monopolies are private. The example of electricity is of greatest significance for natural gas, since in California, across America, and all over the world, most new electrical power stations are fueled by natural gas. The problems that occurred in California were partly homegrown, but the wider problem that America faces now with power stations furloughed or running way below their capacity because of high gas prices proceeding from supply problems will occur across the world in due course, and the money that has been invested in gas-fired power stations will then be understood to have been wasted. The money is being wasted now, however, and now is the time when this folly should be stopped. It won't be, of course, because people need power, particularly electrical power, and at their current levels of usage only big power companies can supply it. In chapter 9, the idea will be developed that the only path to energy independence is to provide energy locally, especially local electricity, but that will mean using a great deal less energy of all kinds.

United States of America

It should already be clear that a sufficient natural gas supply for the United States is no longer possible with current demand. By late spring 2003 the alarm had sounded in government circles. Ostensibly, the trigger for official concern was the price spikes seen early in 2003, the result of high demand from a slightly colder winter than usual, a decline in production from both Canada and the United States, and the consequent heavy drawdown on the gas storage system, close to what some consider danger levels. The natural gas storage system is a very important part of the U.S. supply system. It is a series of large natural underground caverns in various parts of the country that are filled or reinjected during the months when no gas is used for heating. The filling usually starts in April and ends in mid-November. The system has a quoted capacity of about 3.5 Tcf but this is misleadingly low, since it refers to working gas, the part that can be drawn off. The actual amount of gas in the ground is much larger. However, since the 2003 price spikes much attention has been paid to the amount of gas in storage, and when more gas is injected than expected, Wall Street reports on it avidly, and the price of gas drops, at least for a few days. At the end of refilling in both 2002 and 2003, storage stood at higher than the five-year average. Yet the winter of 2002–03, which was not a severe one, saw a record amount of gas withdrawn. With the supply from the United States and Canada falling, the existing storage system can no longer guarantee a full supply for winter heating for anything other than a relatively mild winter.

The calling of the U.S. natural gas summit in June of 2003, attended by industrial users and producers (though noticeably lacking in ordinary users or citizen groups), seemed to be a prudent reaction to a sudden deterioration of gas supply. There is, however, a darker reading of what happened, one more in keeping with the U.S. government's increasingly obvious use of events, apparently accidental or possibly stage-managed to further a preexisting agenda. Although during this highly unusual meeting, lip service was paid by some in the Department of Energy to the normally foreign idea of energy conservation, the main message was a constant drumbeat toward more drilling access to restricted lands and waters. Once the gas supply report by the National Petroleum Council (NPC) appeared three months later,[9] taken together with states suing the administration for relaxing air quality rules to allow more coal burning by power stations, it became apparent that the gas crisis was very convenient for the government of the day.[10] They could now use the evident shortage of supply to try to force through legislation to allow for the streamlining of permits to drill, and try to persuade Congress to mandate a new "inventory" of oil and gas prospects on the eastern and western coasts of the United States and off the western coast of Florida, in the highly productive Gulf of Mexico.

The increase in coal burning, which reduced gas-fired electricity activity, along with the partial destruction of the U.S. domestic fertilizer production industry, appears to have been enough to allow for record injection into the storage system. The record injection was noted with great éclat by the financial press, who previously had not been too interested in gas storage. The question for both the government and the user, especially the home user who cannot easily switch the furnace to burn olive oil or cow dung, is whether there is enough natural gas in storage to make it through the coming winters, especially if they are colder than usual and if Canadian production decline gets steeper[11] and begins seriously to curtail exports to the south. If the United States gets through to April 2004 without major disruptions, it will not mean the problem has gone away, but the question will keep recurring each year thereafter, either until demand is greatly reduced or until a massive number of new LNG terminals come online later in the decade.[12]

The weather will continue to be a large and unpredictable factor in how badly winter gas supplies are stretched, but once new gas-fired electricity plants really begin to flex their demand muscles, they will start consuming a lot more gas, which will have a very strong destabilizing effect on the whole gas market.

The U.S. government will be hoping that the access to new lands it is trying to force through the legislature will raise supply enough to get by until the LNG cavalry arrives—from foreign parts, of course. Its other great hope, Alaskan gas, is at least a very uncertain decade away. Assuming that new drilling in the Rockies or off the Florida coast is achieved, most likely against the strong wishes of the local population, it is still highly questionable whether it will be able to make up for the ever-declining rate of conventional gas extraction. CBM is slow to come online, sometimes taking a year before a well yields useable quantities of gas, and by early 2004 there were reports that one of the most important CBM regions of the country, Powder River, was no longer able to increase supply.[13] New offshore drilling farther out in the Gulf of Mexico is ever less likely to yield the precious gas, and more likely to give oil, thanks to a geological quirk in this area. Much depends on very deep and expensive drilling[14] in shallow water near the coast. In recognition of the costs and difficulties of offshore deep drilling, in January 2004, the U.S. government offered significant royalty reductions to encourage more exploration.[15]

The course outlined above is a considerable gamble, but it is one that other governments should pay close attention to, since they may find themselves in a similar position, namely, short of supply but hooked on gas on the one hand, or else, if they have the misfortune to be sitting on sizeable deposits of gas, they may be on the receiving end of military attention from the world's most aggressive nation. The United States has demonstrated an ever-increasing willingness to attack whomever it pleases, whenever it pleases, for whatever reason it decides to concoct—though it has become obvious to almost everyone that all the new targets have oil, gas, or pipeline possibilities, or in the case of Iraq, all three. There is little any national government can do to stop the United States, since every country on Earth is now also more or less addicted to the same substances, and they too will soon be trying to stop the same industrial system from breaking down for want of ever-increasing amounts of hydrocarbon fuel. In the 1930s, Roosevelt was able to "save" the capitalist system from itself. He had the useful and salutary example of the state of communist Russia to draw on, but even more vitally he could draw on an astonishing and apparently endless new supply of energy in the form of Texas oil. Seventy years later, Texas oil is in its dying days, but those in the world of orthodox economics, which includes a depressingly large percentage of mainstream environmentalists, are hoping that natural gas will perform the same kind of miracle now as oil did after the 1929 crash.

The stakes are thus very high for natural gas, and there is a very large political dimension—which will be discussed more in the following chapters—when considering energy security, mainstream policy, and what else can be done in the absence of meaningful change from the corporate-government complex.

Canada

Some of the same production questions that apply to the United States can be asked of Canada, since it too is now in natural gas decline, but it has two extra and very serious problems. Canada is hoping that CBM will deliver as handsomely as it has done in the United States, and perhaps with less pollution. Much of the coal in Canada, or more particularly in Alberta, since about half its coal is located there, may not be as suitable for methane extraction as that in, say, the Powder River basin in Wyoming. Since CBM is generally slower to come onstream even under ideal conditions, and given that the industry is being very reticent about its CBM experiments, it is not possible to say in early 2004 how much CBM will make up for conventional gas decline in Canada.

If CBM and other unconventional gas cannot make up the deficit, and gas production continues to decline, then some very difficult questions will appear for Canada, thanks to its need for gas to produce synthetic oil from the tar sands of Alberta, and, equally ominously, thanks to Chapter Six of NAFTA.[16]

The tar sands are often claimed to be vital for North American oil independence. Since it is almost impossible that they will ever produce more than 3 to 5 million barrels per day (of what is called synthetic crude oil), the Canadian tar sands can only ever be an auxiliary supply of oil in a continent using 24 million barrels of oil a day.[17] Tar sands contain, as the name suggests, not oil but a tar or bitumen substance that must be mined from the surface or extracted from deeper down with the help of steam, and then processed into an oil that will flow in pipelines and be useable in ordinary refineries. As noted in chapter 3, the tar sands projects have many problems, including damaging the adjacent environment, releasing large quantities of carbon dioxide, and using up vast amounts of scarce, precious water.

Of much greater seriousness to the producers, however, these projects at present all require huge amounts of natural gas.[18] If the tar sands are to deliver on the promise of 3 million barrels a day, then a quarter or even half of all Canadian natural gas will have to be diverted to Fort McMurray, the center of tar sands operations. This will compound the decline in Canadian gas production, and potentially seriously complicate North American plans for reducing foreign oil imports. It will also add fuel to the fire of the problem of NAFTA.

Chapter Six of NAFTA, sometimes called the proportional sharing agreement, was not signed by Mexico. It commits, or condemns, Canada to export its oil and gas south to the United States in the same proportion as an average of the last three years, until the reserves are exhausted. As Canada exports nearly 60 percent of its gas to the United States, that means as its production declines, and assuming that neither the new finds in British Columbia nor CBM arrest the supply fall, Canada will still have to export almost 60 percent. If they comply with this, Canadian industry and citizens will start to find themselves first paying more for gas, and then actually having insufficient gas to meet their needs. More than three-quarters of Canadian homes are now heated with gas, and in a country that in most parts is extremely cold in winter, expensive gas or outright shortages could result in some unpleasant headlines, as people start to die of cold. NAFTA and natural gas could be the noose that hangs Canada, or at least become a political time bomb waiting to go off when the United States presses Canada for its "proportional share" of gas.

The problem may be compounded by a prime minister, such as Paul Martin, who is very keen to keep the United States happy and happens to believe in a similar kind of neoliberal wonderland as the United States does, where everything can be solved by free markets, prices, and endless ingenuity and innovation. None of these things actually runs a furnace in winter or a power station feeding a growing hunger for air conditioners in summer.

Mexico

One of the many puzzles in the gas business is why Mexico started importing gas in the late 1980s, paused briefly in the late 1990s, then began importing in earnest, with no relief in sight according to the EIA.

Elected in 2001, Mexico's president Vicente Fox pledged himself to find enough gas to meet Mexican demand. He included imports but stressed greater domestic production. There is no doubt that Mexico could market more gas than it does. In 2003 it was still flaring much of the gas that came up with its large oil production. Most U.S. energy commentators blamed the inefficiency of Pemex[19] for the poor performance of its gas sector. Whatever the merits of their arguments, they are not unbiased, since there has been a massive push by the United States to persuade the Mexicans to allow investment by foreign private corporations in Mexico's gas and oil industry. As usual, this is not for the charitable benefit of the host country but because the United States is desperate to make Mexico increase its own gas production, and because U.S. petroleum corporations badly need to get their hands on more reserves. These underhanded motives are always masked by free-market language about competitiveness and efficiency. But Mexico's dilemma is very similar to that of the United States, irrespective of how good or efficient its gas industry is. Mexican gas demand is rising fast, in part because it has built a lot of new gas-fired power stations or is converting its older oil-fueled plants. Forecasts suggest that its gas demand will be 9 Bcf per day by 2010, or about one-sixth of the U.S. need in 2003. At the same time, it is hoping to boost its gas production greatly—to 7 Bcf per day.[20] Texans are very much hoping they will see some of that increased production, because their own once large fields are declining fast.

Something will have to give, and the most likely thing is that someone will go without. For several years prior to 2003, many Mexicans had been suffering rolling blackouts because the electric grid couldn't deliver enough power. However, Texas gas producers are actually keener to sell to Mexico than Chicago because they can get the same price and pay less transport costs. Without U.S. government intervention, which usually happens once the free market starts working the "wrong way," Mexico is going to make the natural gas situation worse in the United States.

Europe

Europe is already highly dependent on imported oil and gas, with the exception of the Netherlands, Denmark, and Norway, which is not part of the official European Union.[21] These imports come partly from other oil and gas producing European countries, but natural gas will come increasingly from the same Axis of Gas that everyone else will be turning to.[22] The European gas industry predicts that natural gas use will rise from just under a quarter of all primary energy to nearly 30 percent by 2020.[23] By then, both the Netherlands and Denmark will be nearing the end of their natural gas reserves, and unless Norway has dramatic success in the Barents Sea, it will be in its sunset phase too. Europe will be almost totally reliant on distant and fragile sources of vital natural gas.

China—A Hungry Giant with Deep Pockets

China's huge and fast-growing population is also rapidly becoming industrialized. This is reflected in its economy, which, at least in the early years of the new millennium, is growing at an astonishing, probably dangerous, pace. This growth is causing an astronomical increase in energy consumption, most noticeably in oil for the exploding car market and for natural gas to fuel both new power stations and those converting from coal, and for industrial feedstocks and fertilizers. The classic warning signs have already arrived in China, and though Chinese officials are aware of problems, they are interested in the same kinds of supply-side fixes as the United States and Britain. China began importing oil in 1993 and will soon be a net gas importer too. Chinese oil production is now in decline. It is already competing with the rest of the world for oil from a diminishing number of sources.

With coal dominant, China's gas use has been a relatively small part of its energy picture, at 3 percent of primary energy. But it is still a lot of gas[24] and China's reserves, while considerable,[25] are not going to be anywhere near enough, especially as it plans to double its gas use by 2010, mainly to fuel new power stations. To help supply the extra gas, China expects to start importing gas, as LNG, from a new "local" producer, Australia, in 2005. That will help for some time, but Australia is not Iran or Russia, and it can either make its reserves last, which is highly unlikely, or else it can extract them as fast as possible, as Britain did, and it will be out of the major league in three decades or less. China has been trying to get a pipeline built from Russia, but to date, the way has been anything but smooth, and the Japanese are in hot competition for the same oil and gas. China overtook Japan to become the world's number two oil user and importer in 2003.

China not only has a large population and land mass, it also has large amounts of foreign currency to spare. While it lasts, this currency reserve will enable it to buy what it is short of, though that does not include clean air and water.

India—A Hungry Giant without Deep Pockets

Like China, India is also a population giant with a fast-growing economy, though proportionally slightly smaller than China. However, India has far fewer energy resources than China and is already an oil and coal importer. Its oil imports are set to mushroom, and given that its gas production can barely keep pace with consumption that will see gas use double from 1995 to 2005, India now expects its first gas imports, via LNG, in 2004. India does not have the foreign currency reserves and credit that China does, and its options are more limited. Both China and India, which have nuclear weapons, are interested in increasing their use of nuclear reactors for power generation.

By the end of 2003, no country really seemed to understand the true nature of oil or gas depletion. The U.S. government had certainly been told about both sorts by energy banker Matt Simmons, who is an adviser on the NPC and was on the Bush Energy Transition Team in 2001. Most other nations were busy parroting the pronouncements of the International Energy Agency or the United Nations, which, officially at least, seemed oblivious to what lay ahead. It must surely be one of the most bizarre situations in history and is a testament to the triumph of blind faith over empirical investigation.

7

Energy Security

The Politics of Energy

The final chapters of this book are, of necessity, much more speculative than the previous chapters. The situations addressed here are
some of the most complex and fast evolving, not to say deteriorating,
on the planet. However, these issues must be discussed urgently, even if
it does entail much speculation and interpretation. It is hoped that at
the very least the discussion presented here provides material for careful
consideration. The issues confronted are of the greatest seriousness and
significance. They are also interrelated in ways that are phenomenally
complex and often bizarre and unexpected. To explore energy security
is to realize very soon that energy is connected to everything.

The picture painted so far is not a pretty one. Anyone who investi-
gates energy to any depth soon realizes that the security implications are
simply appalling. Once again we can turn to the United States for
instruction. The industrial system, to which most people in the West
and many beyond are enslaved, is itself clearly a slave to energy, and
energy on a mind-boggling scale. When American oil production

peaked in 1970, the world literally changed, except that nobody knew
it. There was no playing of the Last Post,[1] no funeral anthem, no ago-
nized headlines. With the exception of a few renegade petroleum geologists, no one even noticed. Yet in a sign that that something had indeed
happened, gas lines had already begun in the United States well before
the Arab oil embargo of 1973. As early as August 1971,[2] President
Nixon, already finding that the Vietnam War and related currency
problems were more than the American empire could cope with, introduced forms of gasoline rationing and price controls. Fights among
people waiting to refuel their cars were commonly reported in 1972.

Energy Independence

Energy security, according to President George W. Bush, means energy
independence and economic security.[3] It means avoiding blackouts,
sky-high energy prices, and dependence on foreign energy sources,
while at the same time ensuring a stupendous increase in energy
demand. The U.S. case is in fact quite typical, with predictions of oil
and natural gas use increasing by more than a third in twenty years and
electricity demand by nearly half.

The stream of warning signs that the whole energy supply system
in both the United States and the world is stretched to the limit finally
broke into a flood during the summer of 2003. There were colossal
power blackouts all over the industrialized world, from the biggest
blackout in North American history on August 14 through Europe to
Japan. In fact, though these events hit the headlines around the world,
there had been many smaller but similar events in "advanced industrial" countries in previous years.

Even among the market-driven English-speaking countries, there
is growing realization that energy can no longer be taken for granted.
It is ironic that their blind faith in markets, prices, and ingenuity is

helping to drive them ever further into overshoot, so that it will be somewhere between very hard and impossible to recover. This is not possible to admit or understand if you are trained in economics, but neither energy nor the environment knows or cares anything about economics or ingenuity.

The record of success for governments pursuing energy independence has not been impressive. In fact, unless they have found enormous new deposits of oil and gas, it has been an utter failure. The list of energy independent "developed" countries extends to exactly three: Canada, Britain, and Norway. Britain is about to lose that status, and Canada is not that far behind, thanks to NAFTA. For the rest, many countries, like Brazil, speak of achieving energy independence in this decade. Though some might come close for a short while, it is an absurd and impossible long-term goal for all nations who pursue the same growth malaise. In reality, industrialized and fast-industrializing countries alike ever more resemble Japan, which imports almost all of its primary energy and has done so for decades. Thus, when the rhetorical arguments designed for voter consumption are blown away, energy security boils down to military power. Again, the United States has demonstrated to the world what this means: the endless so-called War on Terror, which translates to illegal attacks on whoever has what the United States needs and cannot get via supposedly legitimate methods of trade. More than anything the War on Terror, which of course breeds ever more terror, means a fight for energy in a world that is going over the brink.

The reason why the normal, business-as-usual method of energy independence keeps failing is by now obvious. We are using too much energy. By having too many people using too much technology, we have outstripped the planet's capacity to deliver its billion-year store of sunlight ever faster to suit an economic system designed for an expanding universe, but not a spherical, finite planet. In the end, the reason why energy independence is impossible can be reduced to the fact that we use more energy than we can reasonably collect from the sun on a daily basis. We are living on credit. In some ways, we have been putting off the day of reckoning for more than 300 years, ever since the unfortunate creation of central banks and the control over nations that they immediately assumed. We have fought more and more wars, always trying to expand the economic system and the markets that they require, and always looking for yet another willing creditor—either a new form of mass energy like oil or natural gas, or a new human resource stock, such as cheap Chinese labor, that can be conquered by military or markets, or some mixture of the two.

When the Japanese wouldn't submit to U.S. trade in 1853 they were met with gunboats. One of the reasons the United States needed Japan was for refueling its merchant fleet.[4] Earlier in the nineteenth century, the British deliberately addicted millions of Chinese to opium, which the British supplied from India, in exchange for tea and other Chinese goods. Chinese resistance to "open markets," particularly an open market in addictive and socially destructive drugs, was met with military attacks by the British in the Opium Wars of 1839 and 1856. In both cases, the Chinese lost and were forced to submit to open trade, not only with Britain, but also with the United States and France. It is a story as old as it is gruesome and brutal. In the twentieth century, addiction to opium and heroin only increased, but it was dwarfed by the global addiction to oil and gas. In the early decades of the twenty-first century, abundant supply of both will certainly run out, with oil going first. It is a sad irony that the blindness of governments, of corporations, of academia, and of the mass-media-manipulated public to the obvious fact that there are limits to physical growth is reducing, with each passing day, the chances that we shall avoid a great and very painful die-off. We are far into overshoot in almost every major system of planetary operation, from potable water to soil, from emissions of global warming gases to horrific toxins that no organism has evolved to deal with. All of this destabilization and destruction has ultimately been made possible by energy use on a huge and unparalleled scale—a scale that now clearly challenges the planet itself, and is obviously starting to overwhelm it. It is equally obvious that the planet will outlast its badly behaved human guests, but the ruin will be great, and much of the planet will be made uninhabitable, at least by many of the larger life forms that have evolved in the last 3.5 billion years. We appear to be engineering another Permian extinction, without the asteroid hit.

Energy independence of the type being proposed by governments and their economic advisors is therefore oxymoronic. Those who have the biggest economic and military muscles will try the hardest to hold onto their highly consumptive modes of existence. The politics of energy is already becoming the politics of everything, and soon it will be obvious that energy security has devolved into economic survival.

The First Blow

Many political relationships already make little sense unless energy is taken fully into account. Until relatively recently, the main energy source to take into account was clearly oil. That is changing fast, and natural gas will soon become an equal source of tension, eventually overtaking oil as the chief cause of strife and wars.

As this happens, it will become increasingly apparent that the natural gas supply chain is much harder to guard and guarantee than the oil supply chain. Without exception, the few suppliers with abundant gas to spare will have to send their gas through huge lengths of very expensive pipelines, often concentrated in hostile and inhospitable terrain. Even Norway is included because its gas starts in the unruly North Sea. Furthermore, either the governments or the people of all the countries of the gas axis[5] are regarded as variously unstable or unfriendly to many Western nations, and particularly to the United States.

The U.S. descent into world pariah status is not going to help it in the coming years of energy scarcity. There may come a time when it is weak enough for the history books to be reopened, when many of its terrible acts will be more openly held against it. The great bully may one day feel the vengeance of the many smaller nations it has tormented. All the great powers relying on vast energy imports will need to consider this phenomenon, which has been repeated many times in history. System collapse does not necessarily happen at the first massive blow, which the body politic may appear to have survived intact. The fall of empires more often happens after several catastrophic events. In retrospect, however, it can usually be seen that the first blow did great internal damage that was not obvious at the time. It is in fact this internal weakening that carries on unrealized and unchecked, further weakening the system and rendering it vulnerable to smaller threats later on. The great blow of the oil embargo of 1973 came along with a suite of disasters that America was bringing on itself, including the dreadful Vietnam War and monetary destabilizations that continue to this day.

Interconnections

As the United States looks abroad for ever more gas, and becomes part of the so-called globalization of gas, it will need to understand things about itself that previously could be ignored, since they were of only regional interest. As in so many other matters, it appears that "When we tug at a single thing in nature, we find it attached to the rest of the world."[6]

This interconnection, indeed overconnection, and the incredible speed of events, particularly political events, is making it quite impossible to understand or keep up with the changes that the energy-political matrix is causing. For example, the arrest of Russia's richest man,

Mikhail Khodorkovsky, who just happened to be head of its largest oil company, Yukos, occurred apparently out of the blue. In fact, the story has many dark and confusing undertones, mostly not reported in the financial press. Soon after the capture of Khodorkovsky, there appeared a string of contradictory news stories, some saying that the new head of Yukos, Semen Kukes, notably a Russian-American, would certainly sell Yukos to the West; others said equally certainly that he wouldn't.

Another much larger example of the difficulty of understanding the mechanisms of oil politics and economics occurred in the 1980s, when prices slumped dramatically and ruinously for many, including both Texas and Russia. Some have called this the third oil shock. Nearly two decades later, there are still highly conflicting interpretations about what happened and why. Some assert that the reduction in oil prices was designed to "take down" the Soviet Union. Others don't even mention the idea. Few now discuss energy without discussing Russia, but equally few know Russia very well, and no one really knows how much oil or gas is under the ground, in Russia or anywhere else. Yet the stakes then and now are extraordinary, and the future of the industrial world depends quite literally on how much oil and natural gas there really is, and how fast we can get at it. It used to be said that Petroconsultants, of Switzerland, had a fairly accurate count on oil and gas reserves, and among their clients, the CIA is probably as well informed as anyone. After its founder died, Petroconsultants was taken over (by the IHS Group in 1996), and many in the industry now have less confidence in this last bastion of technical numbers.[7] The best one can do is to keep a very careful watch on the largest players, both producers and consumers, and try not to imagine that it all makes sense.

Previous chapters have covered those countries with the greatest gas supply and the largest consumers. What follows are some further considerations regarding the strange interconnections and possible political flashpoints. These remarks should be seen against the understanding that during the early decades of the new millennium, those with abundant spare gas supply will eventually shrink down to just the two or three largest producers, namely Russia, Qatar, and theoretically Iran. By perhaps 2030, or possibly much sooner, the world's oil supply situation will be unimaginably desperate, which will only compound the gas picture.

Who Are the World's Largest Gas Importers?

The world's largest importers of gas are presently Japan, Korea, and Taiwan, supplied almost entirely by LNG. Joining the league of giant gas importers, and soon overtaking these three, will be the United States and China. There are also plans for LNG terminals to proliferate in Mexico, partly to alleviate Mexican demand problems, but also to fuel the U.S. demand, both directly via pipeline and also indirectly by generating electricity and sending it over the border. There are even plans for at least three LNG terminals to be built on Canada's Atlantic coast, starting in 2006. If all three Canadian terminals were built, they would have a capacity of 1.75 Bcf per day, about 10 percent of Canada's production.

The LNG plans for Canada should send a shrill warning signal to both Americans and Canadians. To Americans it should say that their exporter of choice is in trouble, and that their own plans for increasing supply will thus fall short, even far short, of expectations. To Canadians it should be a warning to rethink their entire energy policy from top to bottom. Even with business-as-usual and tar sands development proceeding on the present course, there are credible assessments that total Canadian oil production could peak in this decade, despite its enormous tar sands resource.[8] Having peaked in natural gas, and already preparing to become gas importers within five years of peak (very similar to Britain), unless Canada reduces its oil demand and its exports to the United States, it will have insufficient oil for its own voracious needs. During the second decade of the new millennium, Canadians will be faced with the real possibility of becoming net oil and gas importers. Again, this will mirror the U.K. experience.

It appears that what we learn from history is that we don't learn from history. The political effects of this tidal change in energy will be monumental. Canada is not only trapped by NAFTA, but Paul Martin, its new prime minister, is even more free-market and pro-NAFTA than the outgoing Chrétien. Furthermore, just at a time when Canada should be disengaging from the United States, Martin will seek to strengthen ties, which can only mean Canada will be sending more exports, especially energy exports, south. Politicians rarely care much about energy or understand it very deeply, but energy will become a chief driving force in Canada's politics and problems, just as it has been for decades in the United States.

Lessons from Japan?

Japan appears to be an example of a country with virtually no natural energy resources that has defied natural law to become the world's second largest economy. If Japan had to depend on its own energy reserves entirely and could extract them without regard to physical limits, and at present consumption rates, its coal would last just over five years, its gas just under six months, and its oil would be depleted in eleven days.[9] Its nuclear reserves wouldn't last five minutes.[10] Japan is by any estimation dangerously dependent on foreign energy supplies, all of which arrive by sea. Even more remarkably, most of its nuclear plants have been built by the United States, and Japan joined with the United States in 1997 to begin deregulating its utilities. To compound this unwise situation, Japan is entirely dependent on the U.S. Fifth Fleet to protect those supplies through some of the most dangerous and pirate-infested shipping lanes in the world.[11] As energy is seen increasingly as the most prized resource, not just by nations and corporations, but others lower in the "value chain," these supply routes will become ever more vulnerable to attacks by anyone with political bones to pick. LNG tankers will be the easiest targets of all, since they are enormous, immediately identifiable, and have at least the potential to produce an explosion of near nuclear proportions.

This is a snapshot of twenty-first-century energy security—an oxymoron. The world's second largest economy is utterly dependent on the first largest, and in some ways, vice versa, since energy security and economic security are bound together. The force that holds them together is not only military. It includes a vital currency dimension that is also showing signs of severe stress. Because of this, the situation could unravel very quickly. One of the reasons why the United States will protect Japan's energy is that Japan, in turn, protects part of America's ballooning and uncontrollable debt. Since 1971, when Nixon unilaterally took the dollar off the gold standard, this debt had grown to more than $3 trillion by 2003, an amount far in excess of any other national debt. At this level, it may be close to the point where it cannot be serviced, although the International Monetary Fund, being a de facto arm of the U.S. banking system, will be very unlikely to downgrade U.S. creditworthiness to junk status. Nevertheless, the only way such a debt can be maintained and increased is if the credit-granting countries go on buying U.S. dollars and U.S. debt. The rate at which this was happening had reached $50 million an hour by 2003.[12] Thus between Japan and the United States there is a kind of symbiosis at work, albeit a highly pathological and unsustainable one. Just how unsustainable and pathological is shown by increasingly loud calls from Japanese officials, as well as from the population, for major reductions in the 47,000 U.S. forces stationed in Japan.[13] The same officials concede, however, that they need the United States to protect Japan, by which they surely mean Japan's energy supplies. For its part, the United States regards Japan as something of an unsinkable aircraft carrier.

South China Sea

Although the Japanese energy-security situation is an extreme and rather peculiar one now, especially regarding its constrained military power, as time goes on and more states become total energy importers, its position will be less anomalous. For this reason it should be a case study for all nations faced with negative energy supply and demand imbalances. Indeed, South Korea is in many ways in a position similar to Japan, in that it has large numbers of U.S. forces stationed there, protecting energy imports, particularly LNG, but who are disliked both by the Korean population and its leaders.

Once again, the case of the South China Sea is at the same time extreme and worthy of close attention. Its fate affects the whole of south Asia, and thus indirectly both Russia and the Persian Gulf. The South China Sea has already been mentioned as a contentious place because seven countries[14] claim sovereignty over part of it, or in the case of China, all of it. It may have great reserves of recoverable oil and gas. On the other hand, it may not. As in the case of energy in Russia, there are many conflicting reports as to what is happening, with major efforts aimed at resolving ownership disputes in the sea and particularly the tiny Spratly Islands,[15] complete with much aggressive posturing. There have already been outbreaks of shooting, and many standoffs.[16] Any conflagration in the South China Sea would have major consequences for Southeast Asian energy, not just because of the supposed bounty in the sea, but because it would disrupt energy (and other) supplies[17] to many nations, including Japan. Furthermore, at the southern end of the South China Sea, the coasts and countries of Malaysia, Brunei, and Indonesia are already major gas-producing areas. The potential for disruption of production as well as transport cannot be ignored.

Food Security

Another aspect of energy security is food security. Obviously in a globalized food market, large amounts of energy are required to transport food around the world. The number of food miles traveled is staggering. It is commonly noted that an average piece of American food has journeyed over 1,500 miles from soil to stomach. One may fairly argue that, having decided on shifting such absurdly large amounts of food such absurd distances, there are many substitute means of travel and thus of fuel. But there is one aspect of food security for which it is hard to find a substitute, and that is the use of natural gas to make the nitrogen fertilizer on which industrial agriculture depends. Yet again, it is China that will become the most important factor in related world food problems.

Chinese Food

In the mid-1990s, China was warned of the danger of becoming dependent on foreign grain imports.[18] Its harvests and its grain stocks continued to increase[19] until 1998, when they peaked, and have been falling ever since. However, its population has continued to increase, and so have soil erosion, desertification, water depletion and poisoning, more meat eating, and a host of other problems largely resulting from rampant "development" and classic economic overheating. In 2003, serious alarms began sounding in official quarters as the nation's huge grain storage system became depleted to just two years of supply. This dangerous reduction has been caused by a 20- to 30-million-ton grain-production shortfall for more than five consecutive years. As another reduced Chinese harvest came in for 2003, and world production continued to fall, Chinese grain prices began rising steeply in the fall of 2003. China proclaims this to be a positive development, since it helps the market to balance supply and demand, but the situation now may now be sliding out of control.

China illustrates dramatically that the relationships between fossil energy and industrial food production are highly complex. China uses a significant amount of its soon-to-be insufficient natural gas to make fertilizer. At the same time, it wants to increase its natural gas use by a factor of two by 2010, mainly to fuel more electric power stations. It also needs more energy to pump water long distances and irrigate the parched fields of the north. The Three Gorges Dam, which when operating at full bore in 2009[20] will be the largest dam and generator in the world, is also designed to divert water to drought-stricken areas of northern China, the largest grain growing areas. In every direction, China is stretching its energy-food-water system to the farthest limits, and with every new technological "solution" introduces more dependencies on already strained structures.

As a foretaste, during 2003 China was already importing soybeans at inflated prices. If China carries on at this rate, it will start importing 20 or more million tons of grain a year as soon as 2005—that is, if nothing else goes wrong. Those grain imports will help it solve some of its water problems, since growing 1 ton of grain can require as much as 1,000 tons of water, but the large grain exporters are also experiencing water problems and harvest reduction. Those great grain exporters happen to be Canada and the United States, countries now facing natural gas peak. The United States is increasingly importing natural-gas-fed nitrogen fertilizer from such places as the Persian Gulf, which is also eager to export its worsening water problems.

The Rest of the World

What is happening in China is being mirrored at a distance in the other Asian giant, India. In terms of energy security, and most likely in many other security aspects, India and China look set to dominate first southern Asia, and then to pose an increasing challenge to the West.

There is much that is and will be written on the problems of Indonesia, both as a country noted for violence toward its own people and as a giant but declining energy producer. Australia will try to take its place in gas production, but not as an oil producer, since it already imports one-third of the oil it consumes. Natural gas in Africa is creating golden new corporate and government opportunities for expropriation, corruption, war, and impoverishment. That story is already beginning to unfold in the usual bloody and sad way, as the oil story did before it.

In South America, energy security has long been a major issue and severe provocation to all those involved, most particularly the United States, which sees all of the Americas as its sphere of interest and control. The results in once oil-rich Colombia have been and continue to be devastating; in some ways it is fortunate in not having vast gas resources, though what it does have is largely produced by Texaco. Venezuela and Bolivia do have large gas reserves, and both have been targets of U.S. destabilization. In 2003, the Bolivians ousted their Washington-backed president to stop him from exporting their gas via hated Chile to the United States. It may only be a temporary victory. U.S.-backed forces succeeded in removing Venezuelan president Chavez in 2002, but a popular uprising and brave actions by individuals overturned this coup. Ironically, Chavez was quite willing to sell oil and gas to the United States, but he showed intolerable tendencies to disagree with the far-right factions that now dominate U.S. politics.

At the other end of the Americas, the United States faces the difficulty that one day Canada may wake up from its deep slumber and realize that its huge energy reserves have dwindled, and that its nest egg, the tar sands, may remain largely unhatched—to the relief of the many who have to live with the water shortages and environmental disaster that tar sands production causes. In 2004, it has become clear that

Canadian gas production has peaked, and Canada will only be able to maintain its present level of oil production if the tar sands can get enough gas.[21] As explained earlier, Canada already has plans for LNG import terminals on the east coast, and unless it reneges on NAFTA and radically reduces its U.S. exports, Canada could be faced with net oil imports as well. So much for energy security, even when you have vast reserves, if you have a ferocious, "free-trading" neighbor who won't take no for an answer.

Contraction

These are some of biggest problems the world is facing. Almost every country in the world confronts some varying mix of these dangerous issues, if not yet on the same scale. All of these problems are either created or severely exacerbated by our inability and rank unwillingness to control our population growth and our economic growth. In one sense, China was an exception regarding population growth; it tried to do something about it but since the 1980s has been failing, since it expects to add three hundred million people by 2030. Whatever has been achieved in slowing its population growth has been entirely wiped out and reversed by its colossal economic growth, which outpaces everyone else on the planet.

In short, it is clear that there can be no energy security when there is uncontrolled economic and population growth. After global oil peak and North American gas peak, it seems increasingly likely that energy security will not be possible with any economic growth at all. Contraction will be the order not just of the decade, but of the century. The sooner "contraction planning" starts, the better.

8

Where the Hell Are We Going?

C hapter 7 looked at some of the largest political issues of energy at a global level. An encyclopaedia could hardly do justice to the complexity of these issues, let alone a chapter. By contrast, this chapter is shorter because it focuses on energy policy issues, with particular focus on North America.

The Process

Energy, despite being the core of existence and the real driver of the economy, is never a high policy priority for most governments, unless something dramatic happens to interrupt it or make it very expensive. In 2003, the United States became worried about high natural gas prices; then there was a string of huge electrical power blackouts across the industrialized world. Suddenly, for the first time in decades, energy was front-page news. With the impetus of some media attention, mainly negative, the U.S. government finally produced a much heralded and enormous new energy bill. It had been ten years since the last one, and this one had taken two years of secret deliberations and generated much acrimony from those kept out of the process, who happened to be the public. There was even a good deal of disagreement among those inside the process, namely Republicans and big corporations.

In the same year, after a rather more public process, Canada's National Energy Board also produced an energy plan, albeit in a slim volume. These plans, because of the way the United States dominates world energy markets and is deeply dependent on Canadian energy reserves, are highly significant for the globe. They will affect the world for a long time to come.

The U.K. Energy Review came out in February 2003. It clearly has less impact on the world, but it is interesting to note that it admits that energy supply is in trouble in Britain.[1] Still, it offers only the usual technological, ingenuity, and market fixes.

For those in the big energy industries, both the U.S. energy bill and the Canadian energy plan should come as a great relief or even a cause of joyful celebration. Both the U.S. bill and the Canadian plan show a fearless and reckless will to carry on with business as usual at all costs—though only if the costs are to be borne by the public.

Canada: Techno-Vert and the Supply Push

The Canadian energy plan shows little understanding that the western Canadian Sedimentary Basin, which has supplied well over 80 percent of all Canada's gas, has peaked. Unlike the United States, there is little chance of Canada seeing the large coalbed methane and offshore supplement that has helped the United States delay dealing with its

chronic overindulgence in gas. The Canadian plan consists of two separate scenarios—"techno-vert"[2] and "supply push." The first suggests
that some stress be laid on technological efforts to increase efficiency,
so that more can be produced with less—a kind of BAU-lite (business-
as-usual-lite). This scenario delivers about 3 percent annual growth and
sees overall energy use increase by about 35 to 40 percent by 2025. The
supply-push scenario, or undiluted BAU, makes no effort to conserve
or be more efficient and thus predicts energy use to increase even more,
though ironically this will deliver less growth than techno-vert. This is
one reason why even some market fundamentalists and natural capital-
ists espouse green technology: it offers the chance of even more growth,
not less. Both scenarios in the Canadian plan assume that domestic
energy production will go on growing for the next few decades, much
as it has done for the last two. It is a plan of almost surreal disconnec-
tion from reality, offering a choice between growth on stilts or growth
on steroids.

The U.S. Energy Bill

Though much longer and more detailed than the Canadian plan, the
U.S. energy bill offers only one basic path, supply push as a ten-lane
highway, though it does have scruffy little ditches on either side, called
efficiency and renewables. In its own way it is equally surreal, though
the authors, at least those in the White House, had been informed of
the grim situations facing North American gas supply and world oil.[3]

The bill offers about $20 billion worth of tax breaks, subsidies, and
sundry giveaways, almost all of it to big U.S. energy industries and agricultural interests. The primary strategy is to open previously closed federal lands to exploration and extraction of natural gas and oil. The
industry knows that gas is now the real prize, and for years the number
of rigs looking for gas has outnumbered oil at around ten to one. The

American public, however, is focused on oil, since that is already global, and very visibly causing a lot of trouble. The trouble may look new to some, but it is really as old as the oil business itself, and has become much more intense since the U.S. oil peak of 1970. In the United States, the petroleum industry more or less gave up on oil twenty years ago but hasn't ever made a public announcement to that effect.

The second part of "letting the drill bit rip" is the streamlining of drilling-permit applications and easing of environmental restrictions on land. In the important gas-producing area of the Gulf of Mexico, royalty relief has been offered for companies undertaking deepwater exploration.[4] All of this new streamlined drilling assumes that there is a lot of gas to find. At the natural gas crisis summit called by the U.S. Department of Energy in the summer of 2003, one of the most telling remarks was a plaintive call from an independent but long-established oil and gas producer. He ended his speech by calling on the government to allow his industry to go where they wanted, and he pledged that they would try to find the gas.[5] The word "try" should have been pounced on by everyone in the room, but it was not. The industry knows that U.S. gas is going the same way as oil, but they would like a last bite of the taxpayers' wallet before they pack their bags and head for Russia, Africa, and, rather more reluctantly perhaps, the Arctic.

The Arctic reluctance may stem less from fear of the cool temperatures in those parts than from industry worries that they cannot get enough money from the American taxpayer to subsidize the $20-billion pipeline needed to bring Alaskan gas to the Midwest, nor to guarantee a minimum price for the gas transported by that pipeline.[6] Huge U.S. gas-price fluctuations have indeed played havoc with supply in the early years of the twenty-first century, but that is indicative of a declining system left prey to market forces. A minimum price guarantee is heresy to free-market purists, but without it, and enough federal involvement in the pipeline, North Slope Alaskan gas may never reach the lower forty-eight states. The Alaskan pipeline is being sold to labor interests as a huge job-creation package, with suggestions of up to 400,000 mostly unionized jobs being needed.[7] This is a naked attempt to win Democratic support. Since the pipeline won't deliver gas to the United States before 2010, it certainly cannot be claimed that it will make the slightest difference to the gas gulf that has been opening up in North America since the turn of the century.

Worldwide, more than half of all oil and gas is being produced at sea, and the percentage is rising as the easy oil and gas is used up. In a process benignly if rather inelegantly called "inventorying," the production industry in the United States had hoped to open the oceanic coasts up to exploration, after they were closed off in 1988. At least for 2003, it appears that too many coastal politicians, including many Republicans, are worried about tourism and pollution to let that happen. In theory, this safeguards both the Atlantic and Pacific coasts, the eastern section of the Gulf of Mexico, off the Florida panhandle, and Alaska's Bristol Bay, at least until 2012. In the large states of Florida and California the drilling issue, always a tricky one, was particularly sensitive in 2003 with the oncoming election year. Since 1998[8] Florida's governor has been Jeb Bush, George W. Bush's brother, and California was famously won by another Republican, Arnold Schwarzenegger.[9] It is, however, perfectly possible that the stay of drilling off the U.S. coasts is just temporary. When or if Bush is reelected in 2004, the matter could easily be reopened, especially if the United States is in much deeper trouble with gas supplies, as is quite possible, given that new LNG certainly won't be online by then. Even if Democrats, who take only about a quarter of the money Republicans receive from the oil and gas industry, win the White House, that nonetheless amounts to a very great deal of money. If the natural gas crisis grows more severe, Democrats will find themselves doing more or less the same as the Republicans, though for some it will be a bitter pill.

Pork Barrels

These are the more obvious measures designed to increase natural gas supply from underground. The draft energy bill contains many indirect measures, such as exemptions for water pollution and delaying the implementation of air quality laws, which will make CBM cheaper and easier to produce. Such measures will also reduce the amount of gas burned by power stations, not by reducing power demand or improving the efficiency of energy use, but by allowing more coal burning. There are tax breaks to encourage "clean coal technology," but even in the industry few think that it will make much difference before 2010. Hence the need to reduce air pollution controls.

Regarding increasing gas supply, it should be remembered that in the background there is the hope that Canadian gas production won't fall, but in fact will increase, or at least that Canadians will increase their exports to the United States. This is very unlikely, if not impossible, with the most likely future for Canadian gas being a solid and permanent decline.

With the exception of bringing Alaskan gas to Chicago, none of the measures outlined so far will do much for long to stave off the gas crisis.

In what might be called a parody of its own irrelevance, the U.S. energy bill is full of extraordinary subsidies for absurd schemes, the most remarkable being the ethanol section, which forces states to add increasing amounts of corn-produced ethanol to their gasoline to make it burn more cleanly. When you consider the net energy loss for the ethanol fuel it becomes ridiculous.[10] In reality, the only thing this provision really helps are large, industrial agricultural corporations, which have given hundreds of thousands of dollars to the administration in campaign money and are now making millions back on the tax breaks. It is an example of the way energy pork[11] is dressed up to look like energy policy. Another scheme that will do nothing to reduce natural gas use or increase energy independence is the effort to develop a hydrogen economy. While ethanol may be made from or with assistance from natural gas, at the moment hydrogen is almost exclusively made from natural gas. At least in the next few years, any large-scale attempts to switch to hydrogen cars will simply increase the pressure on already tight natural gas supplies. In fact, even industry proponents and the many environmentalists who have espoused hydrogen don't expect it to make much difference before 2010, by which time the United States may have a large and increasing LNG system in place. In other words, rather than importing foreign oil to run gas-guzzling cars,[12] the United States, if world supplies allow, will be importing foreign LNG to run natural-gas guzzlers instead.

In the meantime, instead of trying to increase the fuel efficiency of the U.S. automobile system, which has seen no improvement since the early 1990s, the energy bill contains no fuel efficiency measures whatsoever, other than one paragraph proposing to investigate the possibility of reducing fuel use in cars starting in 2012.[13] This is barely enough even for token or public-relations purposes, especially as the measure doesn't come with a single cent in appropriations. In other words, there is no money to pay for it. The current U.S. administration is either very confident or very certainly tempting providence.

Both as part of the need to increase electricity supply and as a possible way of producing hydrogen for the hydrogen economy as natural gas becomes too expensive or scarce, there are large provisions to increase nuclear power.

While the bill gives fortunes to nonsensical energy-losing schemes such as ethanol made from corn and the hydrogen economy, there are paltry sums for real renewables such as wind and solar, and just $6 million to encourage recycling. That's about 20¢ per head, hardly enough to buy one bicycle tire patch and some rubber cement.

The Coming Energy Crisis

The U.S. draft energy bill represents a landmark of energy policy at its worst. It is almost impossible to see how it could be made any worse or more useless and inappropriate. The United States, and no doubt the world, will pay dearly for this bill, which is the quintessence of highly corrupt pork-barrel politics and payback for all the politicians that the oil and gas industry have bought. For average Americans, this bill is a surefire loser. If the oil and gas companies do find and deliver more supplies, it will only help prolong America's chronic and catastrophic hydrocarbon addiction by continuing the illusion of cheap and plentiful supply. Keeping prices down will make it still harder for renewable energy, largely unsupported by the corrupt subsidies so generously doled out to oil and gas, to make a serious contribution. If the oil and gas companies fail to find and deliver much new supply, especially of gas, then the United States will find itself in a full-blown energy crisis. If the U.S. luck with mild weather (just conceivably a by-product of the global warming it has likely helped to promote) carries on, and if the many LNG terminals are built, then the gas crisis will probably be put off past the 2004 election and then eased by large new influxes of LNG. This could just conceivably happen. If so, then America will carry on as normal for some further years, doing nearly nothing to reduce energy consumption or to develop a strong renewable energy infrastructure.

European nations have made greater efforts to use what energy they have more wisely, but they have obviously had little effect in influencing Washington. It is clear that the engine of economics—business as usual—has no intention of taking any of the bold steps toward massive reduction of energy use that the evidence of hydrocarbon depletion and climate change suggests is so desperately needed. The final chapter offers a different kind of policy from the usual corporate-government effusions, a public policy to be enacted by the public for the public.

9

But What Else Can We Do?

An Unpleasant Talk

Tonight I want to have an unpleasant talk with you about a problem unprecedented in our history." So began Jimmy Carter in a television broadcast to the American people on April 18, 1977. "The energy crisis has not yet overwhelmed us, but it will if we do not act quickly." He then went on to give one of the most prescient, honest, and forthright political speeches ever delivered, and in the light of history one of the saddest and most depressing. It had a reasonable ten-point action plan, which included reducing energy-demand growth (just growth, not absolute demand), reducing gasoline consumption, and trying to keep oil imports to no more than 6 million barrels a day, against a potential of 16 million. By and large, Americans rejected the message, derided the idea of sacrifice, and denied the notion of limits to anything. When Reagan arrived in the White House in 1981, he declared that it was "morning in America" and famously tore down the solar panels from the roof. Ever since then, "sacrifice" has become a taboo word. I was tempted simply to print Carter's speech in its entirety as the last chapter and leave it at that, as a silent and terrible testimony to the slim likelihood that America is going to do any better more than a quarter century later. However, the situation now is even more serious than the one Carter described, and the security situation much more dangerous and pressing. As I will show, those who care must now act on their own and in their communities, because history has irrefutably demonstrated that neither mainstream government leaders nor big business has any intention of acting responsibly or reasonably. I shall make further mention of Carter's historic "sacrifice" speech, and I hope that many people will find and read the speech in its entirety, for it is a remarkable and brave testament.[1]

The Smell of Prosperity

Carter's words, arresting and powerful though they were and still are, came after at least three other very important calls to action in the 1970s, all of which were trampled by the ensuing decades. In 1970, the initial Earth Day, billed as the first U.S. nationwide environmental protest, was designed "to shake up the political establishment and force this issue onto the national agenda."[2] It was undoubtedly very effective at bringing environmental issues into focus and raising awareness that "Americans were slurping leaded gas through massive V-8 sedans. Industry belched out smoke and sludge with little fear of legal consequences or bad press. Air pollution was commonly accepted as the smell of prosperity."[3] In the following years, huge and tireless citizen efforts against corporations and governments did have some success in reducing air pollution, and the oil shocks of the 1970s finally forced auto manufacturers to improve their fuel efficiency, partly by making, or rather importing, smaller, foreign cars. An examination of the U.S. Energy Bill of 2003 and the latest U.S. automobile fuel-efficiency figures demonstrate quite clearly that almost no progress has been made since the early 1990s.[4]

In 1972, there were two events that had the potential to quite literally change the world. In Stockholm, the U.N. Conference on the Human Environment gave form to environmental protection as a governing idea and boldly asserted that there would need to be massive changes in the overconsumptive lifestyles of the wealthy. Stockholm even dared to suggest that large-scale planning, command-and-control methods of regulation should be favored over market allocation.[5] Thirty years later, after a neoliberal onslaught,[6] it seems scarcely believable that anyone would stand up in public and say such villainous and ridiculous things. But there was worse to come. In the same year, the Club of Rome published the infamous *Limits to Growth* report,[7] setting off a firestorm of vicious criticism from just about everyone. Essentially the report stated, in the most carefully argued way, the blindingly obvious: unless we made some very serious readjustment in overconsumption, overproduction, and population growth, we could not continue with business as usual without bringing on the collapse of our social and economic systems.[8]

1970 was the year that U.S. conventional oil production peaked at 9.5 million barrels per day.[9] It never achieved that level again, and more than thirty years later it produces about half that. By the time of Rome and Stockholm, there were already gas lines in the United States, and oil imports were at 35 percent. Then came the oil shock of 1973, when oil prices rose fourfold and consumption noticeably slowed. Jimmy Carter's speech came after the natural gas shocks of the winter of 1976–77. In fact, the only part of his prognosis that was chronologically inaccurate was the global oil peak, which he said would come in the 1980s. In fact, without the two 1970s oil shocks, global oil peak would have hit in the middle 1990s (see figure 6.1) but is actually arriving about ten years later, in the present decade.[10] All his comments made complete sense based on the evidence, and they still do a quarter century later. Even his suggestion that the United States should rely more on coal, because it had such large reserves, makes sense from the perspective of energy security. However, I (and I hope many others) completely disagree with the policy of increasing coal use, because it imposes such severe safety, environmental, and health costs that are also borne far beyond U.S. borders. We are at a critical point in history. How will we provide a new kind of energy, at vast scale, which takes decades to implement? We could choose the path of more non-renewable, polluting, and dangerous energy—that is, coal and nuclear.[11] Or we could take the path of renewable energy and using much less.

The Old New Fuels

The first path shows us that there are lessons to be learned from the two greatest failures of new energy in the last fifty years, namely, nuclear fission and nuclear fusion. How can nuclear fission be called a failure when it supplies 20 percent of U.S. electricity and 75 percent of French, besides being important in many other countries? Well, even after fifty years of development, fission reactors are still not economical. The reactors require huge subsidies in construction, almost invariably go over budget, except in France, where they were built by the state, and the reactor sites require high levels of security for which the state, and thus the taxpayer, is expected to pay (so that that cost is never reflected in the price of operation). And still no one in the world knows what to do to neutralize the radioactive waste. The fast breeder reactors that would have recycled the spent uranium into even more toxic fuel for further fission have been little short of a fiasco, with the breeders in France and Britain having been abandoned amid cost overruns and safety problems.[12] The proposed Yucca Mountain nuclear waste dump is part of a chain of volcanoes and is the third most seismically active area in Nevada. More and more reports suggest that the alloy containers will corrode and leak in less than 1,000 years, long before the radioactivity has appreciably reduced.[13]

If the United States were to supply all its electricity from nuclear reactors, it would need about 500 reactors—that is another 400 more than now, and more than all the reactors on Earth. If the United States were to supply all its energy with nuclear reactors, about 1,000 more reactors would be needed. For the world to do the same, another 4,000 to 5,000 would be needed. The dangers would be unimaginable both of accidents and of even more plutonium proliferation. With such demand, the supply of uranium (or thorium) may not last very long.

Meanwhile, fusion, long touted as being capable of offering energy in almost unlimited quantities with no other waste than very useful helium gas, has had billions of dollars spent on it over the last fifty years, with the result that it is always thirty years away from commercial deployment.[14] In the light of global oil peak (and North American gas peak), we do not have thirty years, and why should we believe the fusionists now, when they have always been wrong in the past? Admittedly, this latter argument is not scientific or perhaps even fair, and it is ironically the same one made by critics of those now proclaiming that we are at global oil peak![15] The serious point of the critique of the nuclear energy solution is that after a very long time and huge efforts and investments, fission is still uneconomic and unacceptably dangerous, and that fusion has not worked even on a small scale, let alone at the monumental level required to replace, or at least relieve, the coming stress.

After four centuries of use, the problems of coal are well known. Coal is harmful and dangerous to mine, both for the miners and the surrounding environment. When burned, coal produces lots of pollutants, which, among other things, cause acid rain, smog, and breathing difficulties, as well as large amounts of climate-changing carbon dioxide.

Solar Rising

Wind and solar are quite different from either of the nuclear examples, but have two problems in a way strangely analogous to fusion. Unlike nuclear fusion, both definitely work, though solar needs to become more efficient in operation and less energy-intensive in its manufacture so that it definitely delivers more energy over its lifetime than it takes to build. It is entirely reasonable to expect that both these obstacles could be overcome within a matter of years, if not sooner. Solar electricity generation is almost mature, and wind is definitely so. The big problem with both if they are to be large-scale substitutes for fossil-fueled power stations is that electricity cannot be easily or efficiently stored at scale. This of course applies to any source of electricity but hurts the widespread deployment of wind and solar because they don't operate well when there is either no sun or no wind. Both in fact depend on the sun, and both should perhaps be regarded as solar generators.

We may be as far from developing a really efficient method of storing electricity as the fusion scientists are from making a reactor that delivers more power than is put into it. The other problem that fusion or any other new substitute energy would also face is the infrastructure problem. It always takes decades (and huge investment, usually in great part by governments) to iron out the problems of systems that have to run with supreme reliability and power the whole nation. Wind and solar don't even provide 1 percent of the world's electricity, and not a tenth of 1 percent in the United States. It is simply unfeasible from an engineering standpoint to expect them to take over even 10 percent of the world's power by 2010, let alone the 50 percent one might hope for. It might have been possible if the United States had not stopped solar development when Reagan and Thatcher brought in "privatization" revolutions. We shall never know, except to point out that Denmark now generates over 15 percent of its power from wind. Even so, it is still highly reliant on traditional nonrenewable forms of power, and critics contend that the grid would not be stable with more than 15 percent of intermittent sources, such as solar or wind.[16]

In the short term, even with great political will and leadership, the world will remain dependent on traditional sources for electricity. Systems more heavily reliant on oil and gas are likely to see price rises, but thanks to the political sensitivity of electricity prices, those nations and regions that can, will do everything they can to keep the price down, by for instance subsidies and relaxing air pollution limits to allow more coal burning. Undoubtedly if the price for electricity and other forms of energy went up dramatically it would have some effect on demand, and possibly prompt more questioning of energy use and supply in general. However, experience in Europe suggests that many people will, if they can, spend what it takes to get it. The main group who would be forced to use less energy are the poor, who cannot choose to spend less on more luxurious items. For the average Western consumer whose life is to a large extent built on easy car access and abundant electricity, moderately higher prices will make no major difference to energy use. This is why I do not believe energy prices are going to be an effective way of changing behavior. By the time prices are high enough, there will be no time left for planning. The rest of this chapter is addressed to those that believe we must start planning and acting now, long before the distorted and mendacious price system finally tells us that meltdown has begun.

Suggestions Both Political and Personal

This final chapter would need to be another book or more likely library of books to come close to offering anything that could be considered comprehensive.[17] As I shall suggest, however, there is no time to wait for that, and anyway, people themselves will have to help write those books even as they explore what might and must be done.

What I offer here are two basic things. The first is an outline of the two linked underlying causes, not only of our problem with natural gas, but with every other resource we are overusing and depleting, namely, uncontrollable growth of population and economy. These causes are too rarely mentioned now, but without them at the center of our thinking, medium- to long-term planning and action will fail ruinously. In the light of these twin drivers, I lay out, necessarily briefly, some long-term ideas for a parallel public infrastructure and some public policies that address overpopulation and runaway economic growth. The ideas are long-term, but the need for them is immediate. In fact they were needed fifty years ago, maybe two hundred and fifty years ago, but that is history.

Secondly, I also offer some small-scale practical steps that ordinary people can take now, personally and in their communities to reduce natural gas consumption particularly, and energy use in general. In case readers feel, like me, that no matter how bad things are, one should start taking action of some sort now, however small and symbolic, I offer a practical suite of responses for high-energy users (which is just about everybody in the West) to reduce their consumption, with the big caveat that they are just a small beginning and must be accompanied by vigorous attempts to begin building a low-energy, parallel public infrastructure. The small steps may seem naïve, but North America is facing a gas crisis with no supply-side fix, and it is facing it now. In fact, if added up, the small steps outlined amount to quite a lot, and without some sudden change in politics as usual, these will take a great deal of will, effort, time, and I am afraid, money.

The events that caused Jimmy Carter to issue his dire warning in the 1970s are designed to show that we have to some extent been here before, and during the 1970s serious efforts were made to change the direction of energy policy. Those efforts were derailed, and deliberately so. Understanding why and how, and what other impediments stand in the path of reasonable action, is vital; otherwise, new efforts will likely meet a similar fate. To help avoid that fate, I must explain the major reasons why nations fail to act responsibly with regard to energy. The reasons can be boiled down to four essentials: economic growth, unreal energy reserve estimates, population growth, and the money system. Population growth is the most unpleasant to contemplate, but the money system is in many ways the worst villain, because it is an entirely human invention born of greed. It is also usually the hardest to understand. However, unless its true nature is appreciated, all efforts at change will be undone by it, just as they will be if population growth is not addressed and reversed.

Sustainable Development

There was a severe reaction to the "Spirit of 72"—the idea (contained in the Club of Rome report and in other places) that we must limit growth. Thirteen years later, in 1987, *Our Common Future*, often known as the Brundtland Report, was unveiled by the U.N. World Commission on Environment and Development. In it there was less talk of command-and-control and much more of market mechanisms. There was no reference to limits to growth. Far from it. In great contrast to Carter, who had made sacrifice the key idea in his speech, Brundtland called repeatedly for "a new era of growth."[18] Endless growth, now renamed sustainable development, is essential according to this scheme: growth must precede, and will be the cause of, any benefits that will accrue to society and the environment. Since 1987, sustainable development has become a central plank in both the work of the United Nations and virtually all green organizations. This is little short of tragic.

Sustainable development is a highly problematic notion. It is essentially an oxymoron, since "development" is a code word for industrial growth, the spread of rapacious free markets, and the forcing of poor nations into being offshore labor camps and cheap producers of export goods wanted by the West, with as few environmental controls as possible. Nearly twenty years later, U.N. Secretary-General Kofi Annan lamented that "in 2002, for the sixth consecutive year, developing countries made a net transfer of financial resources to other countries. Moreover, last year's was the largest such negative resource transfer ever: almost $200 billion."[19] Annan went on to say that "funds that could be promoting investment and growth in developing countries, or building schools and hospitals, are instead being transferred abroad." He said that "such a situation lacked common sense."[20] In fact, the system is working as intended: the rich are taking from the poor. To make it clear that "development" really means "growth," the World Bank uses the two words almost interchangeably, and Brundtland herself did so in her 2000 BBC Reith lecture.[21]

In many ways it is now obvious that sustainable development has become a blueprint for corporate greenwash, public relations, and governmental procrastination. In fact, it was already clear in 1992, at the much vaunted U.N. Conference on Environment and Development in Rio, that the agenda had been hijacked and neutralized. By 1992 it was understood that growth would take precedence over environmental protection, and that the modus operandi was free trade, while market mechanisms were to be favored at all times over the environment or anything else. This was the final blow to any hope of real action on growth. The funeral was held in 2002 at the acrimonious World Summit on Sustainable Development in Johannesburg. There was widespread admission that something had gone dreadfully wrong, but few realized that the problem was much deeper, older, and more malign than the superficial spats that characterized Johannesburg.

Thus, the first reason why the public must act on its own is that with near unanimity all official and orthodox channels support and promote economic growth of one sort or another and espouse policies that have utterly failed, as Annan effectively admits. Endless economic (and population) growth on a finite planet is ultimately the direct cause of all the environmental and energy problems we now face. There is nowhere left to grow: the populations of the industrialized world are saturated with consumption, advertising, toxic food, and endless debt. The poor are either hungry or starving, fighting or diseased, while their countries are deliberately plunged into unpayable debt and ravaged for their resources. China, the apparent exception and last hope for market capitalism, is going into energy debt even as its foreign currency reserves temporarily enter the stratosphere. It is now undergoing the same economic overheating that afflicted Japan, Thailand, and the rest of Southeast Asia.

Suggested Actions on Sustainable Development

We need to have subzero growth,[22] in other words, we need to begin to contract human activities until they match the earth's carrying capacity. By trying some of the following suggestions I hope those of us who are concerned enough will start on the path to contraction, and that our communities will create other initiatives, by thinking hard and critically, and by becoming involved and connected with others.

The Money Fiction

Energy makes the world go round, but it is the monetary system that is the root of enforced growth and an insatiable demand for energy. The massive debts of the poor world are in part a strange result of the oil shocks of the 1970s, in which a flood of Middle East petrodollars was deposited in America and then lent out again at interest by the banks to the poor countries who had previously been relatively unindebted.

The monetary system has so many extraordinary and profound problems that it is not reformable or redeemable in anything like its present form. Unless action is taken to disconnect from the present Western money system, all efforts at reducing energy use, controlling pollution, ending species and habitat loss, contracting the population to within Earth's carrying capacity, or coping with any of the things going wrong on the planet, are simply doomed. These tasks, while not a complete waste of time in the short term, are strategically of no more use than painting the *Titanic* green.

The money that most of us use is a fiction. It is not based on anything real, other than the belief that the United States is too big to be allowed to collapse. The huge and growing U.S. debt[23] is actually what is now holding the whole world-financial system together. It can only go on doing so as long as its main creditors—Japan, China, Taiwan, and Russia—remain willing to allow it. They have to buy more than one billion dollars of American debt every day to keep the lie going. Since they will lose hundreds of billions of dollars if the system collapses, they keep it rolling. For now.

This insane situation is the result of the development, over hundreds of years, of a debt-based, compound-interest-bearing, fractional-reserve-banking system. This means that first, our money is created by debt. Almost none of it is created by our governments, but rather by private banks, over whom we have no democratic control. In the United States, the system is controlled by the Federal Reserve, which is a semiprivate bank.[24] To create more money, banks must make loans. Every time they make a new loan, money is created. Out of nothing! This number money was formerly ink numbers entered in a ledger; now money is just binary digits in a computer.

Certain metals, especially gold, have frequently been used as currency. Gold is heavy and awkward to carry, especially in quantity, and a possible target of thieves. Instead, the rich deposited their excess gold with the goldsmith, who kindly kept the money in a safe place, and gave the rich person a piece of paper with which to reclaim the gold. A tricky problem soon arose with this system, however. The gold often just sat in the safe for years, and no one ever came for it. The owners were trading the bits of paper, rather than the gold itself. The gold-smiths decided to take a chance and lend out the gold in their safe to those who could pay the gold back in time with interest. Some societies have regarded this as a criminal act called usury.

In practical terms, the minor difficulty of being no more than a common thief could be solved by making sure that there was just enough in the safe to pay out if a gold owner should demand his metal back. Trial and a few painful errors led bankers to discover that they should keep about 10 percent in their vaults. Most people didn't really want the gold either; they were quite happy with the bits of paper. So the reality emerged that almost all the gold stayed in the vault, and the goldsmith or bank could lend out about ten times what was originally deposited. The system is called fractional-reserve banking. The bank keeps a fraction of what they have lent in reserve, so that they can pay out when someone comes in and wants all their money, a rare occurrence that has allowed the system to succeed. It also highlights the danger of loss of confidence in the system, either because a foreign army is about to attack your town and you want to get out with all your money, or because you worry about the economy and the value of your currency.

The connection with endless growth may now be clear. If not, just ask, where does all the money come from, both to pay back the original loan, or capital, and the interest? Most of the money, at least nine-tenths of it, that was loaned out never existed in the first place. Even more clearly, not one cent of the interest ever existed. Interest is a complete fiction of the lender, but it has to be paid and thus generated somehow. The recipients of a loan, therefore, can do one of two things: either print money themselves at home, often called forgery, or engage in economic activity, usually doing something that makes a profit.

That profit must be more than all the costs involved in the activity, plus the interest to be paid on top, otherwise the operator will go bankrupt. Notice, however, that it is not usually the bank that is broken, but the debtor.

So the economy must grow to create the profit that will pay back the interest and capital, most of which never existed in the first place. Growth and the money system are locked together. The one forces the other to grow, and vice versa, but the monetary system administered by the banks drives the process. The system must expand; otherwise, it will literally implode, and all the creditors will come rushing for their deposits, just as happened recently in Argentina. To avoid this catastrophe, the money supply must be continually expanded. Once upon a time, kings and governments could print more money, but that requires a lot of work, and the banks realized that creating number or nominal money required no material effort whatsoever.

In the United States and most other industrialized countries banks create more than 95 percent of all money in circulation by creating debt. This is the real reason why we must have endless economic growth. It is not a matter of ideology or reason. It is an artifact of a human-made system, much to the benefit of a few and the disastrous detriment of everyone else, including the poor planet, which, being physically limited, can't cope with a system of eternal expansion.

As with so many other aspects of the story of human beings on Earth in the late industrial era, the whole monetary system itself is coming close to its limits. Specifically, the U.S. debt is now so large that it can only be paid by borrowing more money from the very foreign investors to whom the debt is already owed. There are growing signs that their willingness to carry on doing this is slowing down. With the advent of the euro, there now exists another currency to hold as a foreign currency reserve. Thus, for the first time in decades, there is a workable alternative to the dollar, apart from gold, which also looks as if it has a very healthy future, at least as far as price is concerned. For Americans, the possibility of a monstrous collapse of the dollar, along with their economy and military strength, must be regarded seriously. U.S. military strength and the dollar standard are at present the only two things guaranteeing U.S. energy security.

Suggested Actions on the Money System

Unless we begin seriously disconnecting from the debt-based money system, we shall neither be able to address the problem of endless growth, nor rebuild our local economies, which is the only long-term route to using less energy. Rebuilding local economies and culture on a large scale is the process I have termed "global relocalization." It is imperative that local communities begin to institute a local-money or community-currency system,[25] and form local energy banks, designed to make loans to people (in both national and local currencies) to help them install slow payback, energy-saving or -making devices, such as home insulation or solar panels, and buying land for zero-petroleum food production.[26] Developing a local money system is no small undertaking, but one of the pleasures and benefits of global relocalization, which involves moving from a fuel to a foot economy, is that local people will learn how to do things themselves, work together, and discover their many hidden talents and old-fashioned human resourcefulness.[27]

To reinforce economic relocalization, ask the provenance not the price: in other words, where do things come from, not how much they cost in national currency terms. An obsession with small price differences allows us to become pawns of corporations who can sell things to us for "ten cents less"—they in turn are able to develop monopoly control of everything.

Corporations

As almost anyone from Cochabamba to Calcutta knows, corporations have amassed unparalleled power over governments and ordinary people. They are legal fictions, but their effects are very real. They are designed only to concentrate wealth into the hands of their largest shareholders, to internalize profit and externalize costs—onto the public and the environment. Although they are called "public," in reality corporations are the opposite, being controlled by a secretive, tiny interlocking elite, whose venality and corruption is now slowly being exposed, but thanks to the fact that the governmental and legal systems in most countries have been neutralized by those same corporations,[28] those few executives who are caught generally escape retribution. Corporations have now soaked up so much of the wealth of the world, that even Western populations find that their whole system is constantly desperately short of money, and in debt or threatened with it. Corporations are incompatible with global relocalization, that is to say with a decentralized, post-carbon[29] world, where people have some control over the sources that fulfill their daily needs.

Suggested Actions on Corporations

The key elements that have allowed the development of what Bakan[30] calls the "pathological corporation" are immortality, limited liability, profit maximization, corporate personhood, and shares and dividends.[31] A corporation is pathological by design, Bakan says, because it is legally enjoined to be irresponsible, manipulative, grandiose, lacking in empathy, have asocial tendencies, and engage in deception.[32] The most powerful action that citizens can take is to try to revoke the charter of incorporation of a corporation. This is done all the time by the authorities in the United States, but normally only for small corporations, and for activities such as minor tax irregularities.[33]

Whenever possible do not buy goods from global corporations or from any business that offers shares on a stock exchange. You will be amazed at how incredibly difficult this is. Use locally owned businesses whenever humanly possible. "Public" corporations are sensitive about their image, partly because their share prices can be affected by it. Sell your stocks and shares, and be public about it. Encourage others to do the same. You will be regarded as a lunatic by many for doing this. Read about the history of the 1929 stock market crash.[34] Ask questions about the pension funds that you may be part of, possibly without realizing it. Reduce your debts and get out of debt completely, if you can. There are many organizations working for the reform of corporate legislation and revoking the corporate charter of specific companies. Consider joining them.[35]

Population Growth

As if the situation with the debt-based money system and fanatical economic growth were not hideous enough, it must be mentioned that the other great driver of energy consumption is of course population growth. Population expansion is intuitively rather easier to grasp than the money system, but it is a problem of terrible proportions. Because the world food-production system is now so dependent on hydrocarbon inputs, global oil peak and North American gas peak are going to force food supply problems and thus population issues back into the mainstream.[36]

The dreadful truth will soon emerge that we must reduce the human population to about a billion people or less, to have a chance of feeding them without oil and gas, quite apart from a host of other reasons that have been boiling up for decades and centuries. If each woman had, on average, only one child that reaches maturity—world population would fall to about a billion in 100 years. The only humane and decent way to do this is to reduce fertility voluntarily. If we don't, it looks as if nature will do it for us sometime during this century, much as happened on

Easter Island some 500 years ago.[37] The "natural" method of mass population control will cause untold pain, suffering, and devastation.

There are no democratically elected politicians, with one exception,[38] who have population reduction in their policy portfolio. No government except the Chinese[39] appears to have the semblance of a policy to deal with population growth. There are questions about how it has been implemented and how successful it has been. However, the twin U.N. and nongovernmental organizations (NGO) policies of development leading to demographic transition have obviously not worked. Whatever personal, religious, and political objections people have to population contraction, the problem won't disappear, any more than the energy problem will.

Suggested Actions on Population Growth

Don't wait for nature to bring the human population back into line with Earth's carrying capacity, which is maybe about a billion people: reopen the discussion about human overpopulation and consider serious, democratic population-reduction policies, however uncomfortable that makes us feel. Dispel the myth that only children are social misfits.[40] Limit your personal family size to one child and tell others of your environmental choice. Support organizations that help people, especially women, understand their reproductive choices.

Energy Supply Peak

Many of the physical limits of the planet have now been reached.[41] The earth is telling us that it cannot give up its contents any faster; it has reached a peak and soon will begin to decline. More technology and more drilling can temporarily push the extraction of oil and gas higher in some places, but invariably that makes the decline and fall even faster and more unpredictable. Our air and water were pushed past the limits of their ability to absorb the toxic results of using too much energy long ago. The same may be true of our climate and soil.

To recap what the situation is at the beginning of 2004: we have suffered from unreal energy-reserve estimates since the extraction of hydrocarbons began; however, global oil is in its peak-plateau phase, and the whole world oil-supply system has either no spare capacity or very little with which to avoid shocks, particularly a shock to Saudi Arabia, such as a major hit on a pipeline or oil province. Saudi Arabia has been wracked by explosions on civic targets all through 2003, and its largest oil field may be much closer to exhaustion than almost anyone realizes. Global natural gas supply seems poised for growth, but North American natural gas supply has peaked. The only way the U.S. gas supply can be held steady, let alone be increased, will be by importing ever-increasing quantities of liquefied natural gas from foreign countries. The dollar cost will be astronomical; the political, social, and environmental costs are likely to be very high but will not generally be borne by Americans.

The LNG expansion process has just started but expected to be at full throttle well before 2010. However, because of the limits of the planet, global gas supply may already be in decline by 2020 or much sooner. This is an absurdly short time for such a colossal investment. I am not ignoring the climate situation, but in policy decisions that always takes a back seat to economics. Unfortunately, many of the institutions that do take climate change seriously don't take energy supply peak seriously. The United Nations, in particular, publishes extremely high numbers for global oil and gas reserves, thus leading most in green organizations and civil society to deride the very idea of oil and gas scarcity. For this reason, and because of the way the limits to growth debate was decapitated by those in power, none of the orthodox solutions now on offer, either green, light green, or greenwash, recognizes limits to growth.[42] In fact, they explicitly say the opposite, that we must have growth, in the form of sustainable development, to get us out of the mess growth has got us into.

The "walking worried," those who actually believe we are in a very serious situation and want to do something about it, do not have a large number of options, certainly not as many as Jimmy Carter had nearly thirty years ago. Besides, most of us are not the president of the United States, though if he (or she) happens to read this book, there are still a few things those in power could do. Be that as it may, by a bitter twist of irony, the most lasting energy policy legacy from the heady 1970s was from Jimmy Carter. Sadly, it was not his many environmental initiatives, but his unfortunate militarization of energy, known as the Carter Doctrine, which is now the controlling idea in energy policy. Just before he left office, battered by the Iranian hostage siege and the second oil shock following the removal of the hated U.S.-backed shah of Iran, Carter said: "An attempt by an outside force to gain control of the Persian Gulf region will be regarded as an assault on the vital interests of the United States of America, and such assault will be repelled by any means necessary, including military force."[43]

In the succeeding quarter century, military force from the United States and its client states has become the norm. The world, and particularly the poor people of the Middle East, are still living with the ever-worsening results of this dreadful mistake. As the United States looks to the world for natural gas, once again it is the Middle East (along with Russia) that has the largest reserves. No crystal ball is necessary to guess what is likely to happen.

How to Start Cutting Natural Gas Use

Reducing natural gas use should be done with the understanding that such reductions must take place within a larger context of overall energy reduction, and attempts to build a low-energy, post-carbon infrastructure, without which the really large and long-term reductions in energy use will be impossible. Gas, being integral to our system and mostly hidden from our gaze, makes it harder to track in use.

I have targeted four main areas for reduction: heating, plastic, power, and food. This is not a complete list but would make a good beginning. Reducing gas use in these areas will in most cases not be simple and will require careful study of local conditions.

Heating

For those in cold places who heat with gas, most especially in North America, the situation may soon become extremely serious. There is a very real possibility that well before the end of the decade people will freeze to death in their homes because they are unable to pay for gas.

There are two main strategies available for staying warm on an energy budget. No matter what is used to heat a house, it should be very well insulated. Aim for the highest R-value (thermal resistance) insulation you can attain. In many cases this is likely to involve considerable cost, but unless it is done, a large amount of energy will be simply wasted—which will also be expensive. The difference is that one expense comes as a large initial cost, and the other comes over a long period of time. This is a catch-22 for many people and would be greatly helped by government grants and lending schemes that recognize the problem—such as the local energy banks mentioned earlier. Modern industrial insulation uses some oil and gas in its manufacture. There are alternatives to this type of insulation, but they are not yet widely available.[44] For those with Internet access, there will be more and more sites that can help.

Windows are usually the easiest areas to improve and can make a very large initial difference. Double-glazed windows can reduce heat loss through windows by up to 50 percent.[45] Although glass takes a lot of energy to make, if the windows are well made and fitted, they will save a great deal more energy in the long run.

Much better insulation is the first step. But for most it will still be necessary to have a source of heat.[46]

Alternatives to heating with gas should be urgently investigated. The book *Natural Home Heating: The Complete Guide to Renewable Energy Options* by Greg Pahl[47] examines an extremely wide range of possibilities. *Home Power* magazine is another very useful resource.[48] The old-fashioned way in many parts of Europe and elsewhere was to close some rooms down for the winter, and heat (and insulate) only the most important rooms used for daily living. This is not an option for those in houses with few interior walls or doors.

To begin with, the easiest thing to do is turn down the thermostat a few degrees in winter. This is the time for Carter cardigans and warm slippers. In summer, reduce the amount of air conditioning. If possible, switch it off. This is no joke in many parts of North America (and elsewhere), but air conditioning will become both more of a burden on a stretched electricity grid, and more expensive. Much of the rest of the world does not have the luxury of air conditioning, but many would like it. They are likely to be disappointed in their desire. Rather, where possible, buildings should be designed and built to stay "passively" cool in summer and warm in winter.[49]

Plastic

In the industrialized world, almost everything we do or touch or use is derived from oil and gas. Most of the time one cannot really separate the two, but electricity generation now relies far more on gas than oil, and many plastics are increasingly made from natural gas. Therefore, it is imperative that we find ways of reducing the amount of plastic and electricity that we use. There are exercises for use in schools that ask students to imagine what life would be like without oil. There are fewer such exercises for natural gas, but chapter 4 gives many reasons why such exercises should now be undertaken, especially in North America and Britain.

One of the most obviously wasteful and unnecessary uses of plastic is in packaging. However, reducing packaging will entail many other changes, including not transporting so many things so far. The benefits of reducing packaging, or doing away with it altogether (which is quite possible), would be simply astonishing. However, so much of the economy depends directly or indirectly on it, that long-range planning would have to be done, or many people would end up on the streets for lack of jobs. Most of our industrial food delivery system would fall apart too.

A simple way to start is to get cloth shopping bags to take with you when you go out and refuse plastic (and paper) bags. This has reduced the amount of plastic used in Ireland.[50] More dramatically, one may refuse to buy items that have any packaging at all. To put it mildly, this will require a good deal of planning! Almost everything in the industrialised world is packaged. It will almost certainly require items to be locally made or locally grown. However, this will become a necessity, rather than a bizarre rarity, sooner than might be imagined.

Power

Reducing electricity use may sound easier than reducing some other things, but in fact our dependence on electricity is absolute. This is true from California to China, as both have discovered during blackouts in the last few years. However, there is one incredibly simple measure that most North Americans could and should take now: switch appliances off when not in use. Even supposedly highly aware people seem quite unaware that devices not in use can and should be switched off completely (don't leave them in the stand-by mode). For example, desktop computers consume the equivalent of about two or three 100-watt light bulbs, and in the United States computers require about five to ten large power stations to run them alone. There are now about 200 million computers in North America; by 2005 it is estimated that there will be more than a billion computers on the planet.[51] They should be switched off when not in use, or at least put into sleep mode. If you can afford them, buy more efficient light bulbs and turn the lights out as you leave the room.

With both North America and the world turning to gas to make electricity, "switching off" is a simple way of getting used to "power-down."[52] Not owning or using the appliance in the first place is even better, since it saves on the electricity (and other materials) needed to make it as well. It may sound draconian and will undoubtedly require careful planning so as not to cause an economic implosion and wholesale unemployment, but doing without, possibly involving inconvenience and sacrifice, will one day have to become fashionable—again. In the interim, at least having smaller appliances, as is common in Europe will help, though special care must be taken with the discarded item.

Food

It is said that modern industrial agriculture is the use of land to turn oil and gas into food.[53] Since at least the late nineteenth century we have been using more nitrogen fertilizer than the biosphere had to offer. The Haber-Bosch process developed in 1909 appeared to solve the problem—by using natural gas. Today there is no shortage of natural gas for the planet as a whole. Not yet. The United States can and does import nitrogen fertilizer, and in ever greater quantities. For this and many other reasons, North Americans (and all other human beings too) should work vigorously to reduce and preferably eradicate the use of nitrogen fertilizer.[54] There are two grave implications of this statement: there is little question that in practice, we won't be able to produce nearly as much food without natural gas. With oil peak in this decade, and global gas peak following soon after, the leviathan industrial food system must be dismantled. It takes years and decades to change food systems, to rescue and rehabilitate ruined soil. Without huge inputs of (oil and) natural gas, the planet cannot produce enough food for six, seven, eight billion people; possibly not even enough for two billion. We have too many people on the planet. That was probably true a hundred years ago, when there were less than two billion people.

North Americans (and to a lesser extent Europeans) consume up to one hundred times more resources in general than people in poorer countries, and our food consumption is an important part of the reason why. Thus when a North American reduces consumption, it makes an enormous difference. There are also powerful economic and cultural multiplier effects that increase the effect of this reduction. We can start eating lower down the food web—less meat (and fish, which requires so much diesel to catch), less processed and packaged food, more locally grown food, especially vegetables and fruit, grown with no gas or oil. Corporate ownership especially will destroy any chance of keeping local control over local food production. A plant-centered diet uses less energy to supply a given amount of caloric value. Vegetables, nuts, beans, grains, roots, and fruits are much easier to store than meat and fish and do not require refrigeration. Refrigerators are often the single largest user of electricity in an average home.

It amounts to this: reduce or remove flesh from your diet as much as possible.[55] Fashionable high-protein diets are incompatible with this. Reduce or eradicate your use of processed food, and of supermarkets, which are all linked into the fossil hydrocarbon chain of food production. Many supermarkets are owned by global corporations. Form or strongly encourage farmers' markets and organic farmers, and help those who are shifting to organics. It's very tough financially during the early years. Plant an edible garden, even if it's window boxes. Join or form community-based farms. Learn how to can, pickle, and dry food for the winter. Learn how to compost, then do it! Eat and drink together with others, especially homemade food, when possible—this is truly one of life's joys.

Two Problematic Paths: LNG and the Hydrogen Economy

For their own sake and that of the poorer populations of the world, especially those that have significant natural gas deposits, North Americans must resist the call to build new LNG terminals. A new terminal was blocked in California in 2003,[56] but in the future the pressure will be immense. There at least two dozen on the way. Allowing them to be built will expose the world to an expansion of the oil wars that have already caused so much death and destruction to so many, including Americans. Even those who disagree with this assessment cannot deny that investing in the huge and expensive new infrastructure necessary to import liquefied natural gas will certainly condemn the United States to ever more dependence on foreign sources of energy. The one thing that it will by definition never, ever do is offer energy independence. The only way to energy independence is for all of us to cut our demand for energy drastically and try to live more nearly within the quite generous ration provided daily by the sun.

Another large-scale, business-as-usual solution that has become fashionable is the hydrogen economy. As presently conceived, it is largely designed to keep the great car economy running once oil becomes scarce and expensive. Developing this kind of hydrogen system will be a disastrous mistake.[57] Hydrogen is a not a source of energy on planet earth; it is just a carrier, and for most purposes, not a very good one at that. Hydrogen is only a source of energy in the Sun, which is largely made of free hydrogen. On Earth, hydrogen must be made from something else, most often, none other than natural gas. It can also be made in other ways, including from coal and electrolysis of water. On Earth, making hydrogen, and especially storing it at high pressure or low temperature, always requires energy, so that hydrogen ends up as an energy loser.

There may be some application for hydrogen as a storage device for windmills and large solar arrays. Certainly a good storage system for electricity is desperately needed, but after well over a hundred years, we are nowhere close to having such a system. A huge, worldwide public research program is needed to develop better local electrical storage systems. This should be a fundamental aim for university and government research teams. If national governments fail to do this, local and regional communities, including municipalities, counties, states, and provinces, should fund what research they can and disseminate the results as publicly as possible.

Further Suggested Actions on Energy Supply Peak

Be less selfish: even though sharing takes a lot of effort, being selfish takes a lot of energy. Start with schools and universities and other learning institutions. Carry out energy audits; find out where your community uses the most energy, then develop energy budgets. Consider classes in energy reduction, how to run a local currency, and growing oil-and-gas-free food. Those with experimental inclinations can consider setting up a whole money-energy-food system.[58] Reinforce and encourage those who are trying to use less energy. Form local renewable energy groups and work out local problems together. Vigorously support research and development into how to store energy, especially electrical energy, and try to get local politicians involved. Everyone should learn about the laws of thermodynamics to have a better fundamental understanding of what energy is.[59]

Citizen Action

It is true that all forms of orthodox power, including most nongovernmental organizations (NGOs), and the United Nations itself, have almost completely failed to take any of the large steps necessary to confront the idea of oil and gas peak (or population). Nor have they been effective in addressing the 60 to 90 percent reductions in carbon dioxide emissions from burning fossil fuels that would be needed to tackle climate change as if it were a real issue, instead of some optional extra, to be afforded when good times allowed. Looking across a vast range of data now available, it is clear that in almost every area imaginable and measurable the world is going flat-out in the wrong direction. Nibbling at the edges of problems hasn't worked, and it will be ever more obvious that it is a waste of time, effort, and energy. But please don't throw up your hands in despair, and conclude that there is nothing to be done.

Because of the failure of our institutions to act, it now falls to the public, at the level of community and citizen, to take the kinds of action needed. The issue of natural gas peak in the United States, plus expansion of gas production and LNG worldwide, is one of the hardest for anyone to face directly. However, energy connects everything, and all forms of energy are connected. We must, therefore, take action on energy in general. And we can, but it means, as I have tried to show, casting out many of our cherished ideas and embracing some things that appear to have nothing to do with energy, or with many other traditional environmental, social, political, or justice issues.

This is much closer to what Jimmy Carter was trying desperately to say in 1977 and very close to the sabotaged spirit of Rome in 1972. It includes the dreaded "S" word—"sacrifice." Those of us who benefit from, or at least live in, the rich, industrialized world are going to have to start giving some things up. Permanently. Other things will have to be greatly reduced. Or else we can wait and see what happens as we go deeper into the carbon chasm.

A Post-Carbon World

Moving away from our high-energy, industrial system would bring us closer to how a future that still has human beings in any numbers would look, a future that one day will contain very few nonrenewable hydrocarbons: in short, a post-carbon world.[60] There will still be plenty of carbon around, since all life is made of it,[61] and whatever the state of oil and gas reserves may be, coal is still prevalent. But unless the climate scenario changes dramatically, we should be trying to leave that coal underground. In fact, the same goes for oil and gas, but only a fairy story could pretend that we aren't going to let the drill bit rip everywhere on Earth, unless we run out of money first.

The numbers seem to tell an incredible story, and they tell it with more power and resonance with every passing day. They tell us to use much less; much less of everything, but particularly, much less energy, what Richard Heinberg calls "powerdown." Given our current knowledge of climate change and hydrocarbon reserves, the best policy in an ideal world would be to manage without hydrocarbons at all, to create in effect a "post-carbon" world.

Unfortunately, thanks to the laws of physics and chemistry, hydrocarbons are the most efficient and compact way that exists to store energy. In small quantities they are more or less manageable; in large quantities they are plainly a catastrophe. In recognition that what we need desperately are workable, small-scale operations that citizens and communities can act on, some renewable hydrocarbons will be necessary. I suspect that as long as there is humanity on the planet, they will be using some hydrocarbons. We should, nevertheless, aim for post-carbon energy systems—no hydrocarbon use wherever possible, and renewable hydrocarbons where absolutely necessary, especially if it helps make a foreseeable transition to post-carbon systems.

Once we accept that we are entering a post-carbon age, that the Industrial Revolution must in many respects be reversed, the greatest challenges we face are energy storage and socioeconomic infrastructure problems. Many places can get a reasonable amount of solar energy in the daytime in summer. The wind will blow from time to time, and trees will grow. But what happens to the sun in the evening and what happens to the whole energy picture in winter are another matter. We shall have to reduce our energy usage at night and in winter. This is what we did in the past. It is also what many animals do. This is obviously going to be difficult since night is dark and we need light, or at least we are used to it, and winter is cold, and we naked apes shall freeze without some means of guaranteeing external warmth. These are problems of the greatest magnitude, and we may have to face them much more quickly than we realize.

The population problem is also an example of an infrastructure problem. Already we clearly don't have enough food in the right places to feed our more than six billion mouths, and soon we won't have enough food outright. Nor do we have the right institutions or ideas to plan how to reduce our number in a decent fashion, with many continuing to believe that only economic growth will reduce fertility. A critical analysis of the data shows that perceived increasing economic opportunity usually increases fertility. The key is perception, and usually overoptimistic estimates of what wealth and good fortune in life will bring. This highlights the problem of intellectual infrastructure—there are too few thinkers, academic departments, and institutions working to consider what might happen if our industrial system should for some reason fail catastrophically.

A Call for a Post-Carbon World

A post-carbon world will be an ideological paradigm wrench rather than a shift, and all of us will therefore need to greatly enhance our critical thinking skills. Learn as much about global and local history as you can, and not just the official version, which contains large amounts of propaganda and is designed to be mythical in the pejorative sense. Learn how your country has used military force, violence, and genocide to achieve power. This applies to all North American, European, and some Asian countries, not just the United States and Britain. Learn about the dreadful history of capitalism (and communism is no better) and how it cannot operate without high levels of violence and coercion. Discuss new political systems that are not based on capitalism or communism, or any of the post-Enlightenment systems of materialism, all of which require endless growth. Engaged citizens will need to form their own think tanks and action committees to examine what each locale and community might do on its own and with others. There will need to be a harsh examination of the roots and development of liberalism and libertarianism, and of the contradictions between freedom, equality, and democracy. Ideally, however unlikely as it may seem, universities would, as a first small step, set up departments to examine the social, political, environmental, and economic implications of oil and gas peak, both in North America and the world.

Energy Literacy

The extraordinary physical and philosophical infrastructure problems that oil and gas peak will highlight are emblematic and symptomatic of all our problems, especially in regard to energy and how we have built an industrialized world that cannot function without vast quantities of cheap energy. If you go on an energy fast, even for just a few hours, during the daytime, you will find that life is unworkable in most places in the industrialized world.

It is that energy-intensive infrastructure that offers few choices or possibilities to those who wish to use much less energy and consume much less stuff. If you live in most North American cities, and have no garden, and want to stop using your car, you will most likely be faced with long walks to reach the places where you can purchase the items you need to stay alive, starting with food. You will almost certainly not be able to choose a large percentage of your food from local farmers,[62] nor will the store you buy it from be locally owned.

You can dramatically increase your energy literacy, by learning, for example, how much energy you use in your household appliances, how much energy is embodied in the things your use everyday, such as computers, clothes, and food. Increase political literacy. What do the candidates that you can vote for think about these issues? How can you put energy issues on their agenda?

Countering Globalization

English is the language of business, science, and technology, the language of modern empire and pseudocolonialism, and the language of Hollywood, of mechanized, corporate culture and consumption. Most of the ideas that are so problematic came from Britain and were added to and rapidly developed in the United States. The "West" is the biggest per capita energy user, and all parts are complicit in the system of Americanization that is more often known by the public-relations euphemism, "globalization." I don't wish solely to indict Anglophone countries, since Europe and much of Asia is doing the same as the United States, but if systemic and structural change is to take place and have a chance of spreading and being taken notice of, then it must happen in North America and Britain and other English-speaking countries first.

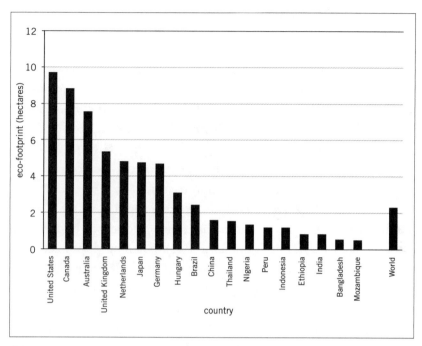

Figure 9.1. Per capita ecological footprints of selected countries. *Source: Worldwide Fund for Nature 2002.* 1 hectare = 2.47 acres.

Lest I be accused of being anti-Anglophone, a glance at figure 9.1 will furnish a further and powerful physical reason why I insist that English-speaking countries are so problematic, and why they must act first. It is surely striking that Americans and Canadians have the largest eco-footprint,[63] on average almost five times larger than the world average, followed by Australians, then by the British. I believe this is not a coincidence. When English speakers lead in reducing their energy use, many others are more likely to follow. Even if they don't, it is clear that this group can and must make the biggest reductions.

Irrespective of language, it amounts to this: to move to low-energy living, or a post-carbon system, we shall have to disconnect from much of what has been built by industrialization and build a new infrastructure that will help us carry on a reasonably decent local life, using a fraction of the energy we now use. This is why the effort to use much less energy cannot be a lone, maverick action. This book is called *High Noon for Natural Gas* because it seems that a host of malign forces has now lined up to make the future not just troubling, but unworkable without major changes. Until our governments accept that we are moving into a world challenged for energy (as well as by many other biosphere problems), the new infrastructure will have to be a "parallel public infrastructure," with the idea that one day it could help replace the decripit, corrupt, and counter-productive institutions we now have in place.

In the film *High Noon*, the protagonist, played by Gary Cooper, confronts some problems, in this case some "bad guys," on his own. Even though Cooper's decision not to run away from his problems is a good idea, the way he eventually deals with them is emphatically not how we should deal with the mess we are in. Though hopelessly out-numbered and injured by one of the cowardly people who was sup-posed to help him, Cooper ends up shooting his problems. Violence is, of course, exactly the method that big power uses to "solve" problems, as countless examples show. Besides avoiding violence as a way to solve problems, the other major difference is that the townspeople—that is the public, you and me—must be convinced to act together, even if not in perfect unison or harmony. While we may not like to think of our-selves as "bad guys," who among us is entirely good? But we can't wait for the perfect "good" person, nor for the perfecting of humanity. Infrastructure cannot be built by individuals; it requires a long-term collective effort. To be able to use less energy, whether it is natural gas, oil, coal, or electricity will mean a lot of collective work.

A Call to Relocalize

We shall have to remake our towns and cities so that they work with little or no use of cars, and it will have to be done widely, the policy I have called Global Relocalization. It will entail making, growing, or getting our daily needs locally, from local people and local producers. Local means not owned by a large, distant corporation. It means not part of a large chain. It means locally owned. This is absolutely vital. We shall need to make the places where we live, work, and study beautiful, so that they are worth caring about and being in. There are so many laws and regulations that help force you into a fuel-wasting, car dependent existence that it is hard to know where to begin, but ending single-use zoning and promoting integrated communities is a good place. As we move back to a foot economy many of the things we shall need to build or rebuild will become obvious. Just how much we have lost will become terribly and painfully obvious. Unpleasant realities must, however, be faced.

Using Less Energy in General

The single, fastest way to reduce general energy consumption is by sharing. It's that simple, and that hard. Sharing is precisely the opposite of what the dominant Anglo-Western culture teaches the world. The individual uses far more of everything than a group that shares, which may be good for the economy, but disastrous for everything else. Whatever the political system, if this obsession with self and the individual is not greatly reduced, our desire for energy will certainly continue to grow. The results of that growth will not remain a mystery for very much longer.

A simple analysis of where we use most energy quickly reveals that private personal mobility—namely, cars—are the largest single item. After that comes electricity and heating. Taking action to reduce very significantly, say by a factor of ten, the amount of electricity or heating that an average home uses is a very formidable, medium- to long-term task. Massively reducing the amount of energy we consume in our daily mobility is, however, a far, far simpler and quicker task. First, we simply aim to use our legs as the primary mode of transport, and we keep that aim for the rest of our lives. That means walking, something that most people can do in most kinds of weather for most of their lives. Bicycling is secondary to walking, but it is a very important and useful second. It is possibly the most energy-efficient form of transport available.

Now come the excuses and reasons for not using our legs, and jolly good reasons they may be. The trouble is that when several hundred million relatively wealthy people all make good excuses, nothing changes. So we need some transitions, and we need them fast. In some English-speaking cities there is a semblance of public transport. None of them matches an average European city, and that is a very important point in itself. For most places, however, public transport is either non-existent or pathetic. For the short term therefore, in most North American cities, there is really only one choice: some form of car pooling on a grand scale. It can and should be done, both by forming or expanding car cooperatives and by instituting formal ride-sharing, or what the Germans rather catchily call *Mitfahrgelegenheit*—"with journey opportunity." In rural areas, these measures may be quite impractical, then again they may not. But the biggest users of energy are city dwellers, so it makes sense to start from there. There is a very large and thriving car co-op in Vancouver.[64] At the end of 2003, there were thirteen hundred active members and almost seventy cars.[65] That works out to one car for just less than twenty people. This is a really serious reduction, and it is being done in an averagely overgrown, badly designed, car-choked, strip-malled, grid-bound, single-zoned North American city with some of the worst social problems in the industrialized world. If Vancouver can do it, practically any city can do it.

Ride sharing works very well in Germany and Austria, and also in some places in North America. It is much more difficult in North America, if only because of the ingrained notion of individual freedom, not to mention fear of violence and mistrust of strangers. Nonetheless, it will have to be tried on a wider scale.

The first result of car co-ops and ride sharing is that easy, spontaneous, medium- or long-distance mobility is definitely impaired. We have to think before going a few miles, plan ahead a bit. We have to walk or cycle a few blocks to get the car. It forces us to examine how necessary that motorized journey is, and whether the job could be done some other way, such as by walking.

Moving from a Fuel to a Foot Economy

A simple catchphrase, but it is the greatest task that industrial humanity has ever faced—getting back on our feet. Only when we start regularly walking to fulfill our daily needs, will we truly realize the awfulness of the autocentric cities and towns that we have invested billions and trillions of dollars in and spent millions of person-years of time building. As we reluctantly get out of our cars and get back on our feet, we can start to see what we need to do to rebuild our transport and "roofed" infrastructure, the places where we live our daily lives.

On foot we can see the sad traces of all the locally owned small shops that have been all but obliterated since the Second World War. We may hear stories of the artisans and craftspeople whose businesses have been destroyed by decades of globalization, so that now when we need a pair of scissors or a shirt, we can only buy them from an insanely large and inhuman superstore whose profits immediately leave for a head office far away. We find, almost invariably, that the item we buy has been made thousands of miles away, most likely in China, where we know that working conditions are often appalling, factory accidents are a dime a dozen, and where nine of the top ten most polluted cities on Earth are located. Of course, most of us have no idea how to make a pair of scissors or a shirt. Our predecessors commonly knew how to do the latter and may have had some clue about the former.

Getting back on our feet will eventually change everything. It will also cause and be helped by a process of global relocalization. In other words, across the globe, we should be striving for an integrated, local self-sufficiency. Global relocalization recognizes that it is not possible for human beings to be entirely, or even anywhere near, fully self-sufficient at a local level. It is, however, a good aim, just as the goal of not using any hydrocarbons is the one that should guide us, even if we don't reach it very often, especially in the earlier stages.

For those who are already members of co-ops, promote the idea as much as you can. This will be true for every other response we find that helps reduce energy. If you are doing it already, start to share the idea. Vigorously. For too long, too many of us have felt isolated and lonely, and, quite frankly, like freaks. Ever since the devastating rise to power of the public relations industry, those who have not wanted to take part in the consumerism that is helping devastate the planet have been ridiculed and belittled, ostracized and neutralized. That time of misery is coming to an end.

A Call for the Foot Economy

Get your daily needs within walking distance of where you live. Live close to where you work or study and close to those you love and those who care about you. Advocate for better pedestrian, cyclist, and transit amenities. Sell your car. If you can't manage without a car, join or start a car co-op. Add biodiesel cars to your car co-op, either from the start if possible, or as soon as possible. Form a biodiesel co-op or make it yourself, if you have some space, even though biodiesel will be a limited substitute, and should where possible be reserved for car co-ops, public transit, and farmers. Biodiesel is no magic bullet: the vegetable oil it is made from requires scarce soil (though multicropping and using the nonfood parts of crops may help a lot), and it gives off pollutants when it burns. Waste vegetable oil, the most common oil ingredient of biodiesel, most likely contains genetically modified soy or canola beans, produced by petrochemical industrial agricultural and biotech. The other main ingredient of biodiesel, methanol, is made from natural gas.[66]

Culture and Politics

Watch less or even no television. This will save quite a lot of energy over the lifetime of a TV set, but the real reason is that television is the main means by which a consumerist and deeply wasteful way of life is indoctrinated into the population. The most damaging effects of television are on children and on democracy, though nothing is left unscathed by it. Giving up or reducing television usage for most people will be as hard as eating less meat or doing without a private car. Television viewing makes the body and mind passive and inert. Political ideas have been reduced to sound bites and culture to the lowest common denominator of sex and violence, because these sell.

Politics is regarded as a very low form of life by many, and government is cursed by even more. But complex societies need governance, and democratic politics has at least the hope of offering justice and help to the less privileged, whereas free-market, hypercompetitive, unregulated, "pathological" corporations have no responsibility to society or the planet whatsoever. Part of building a parallel public infrastructure is to rebuild our political system, and permanently reduce the power of corporations. Ideally to nil. If we don't do it, who will?

A Call for Handmade Culture and Local Politics

Reduce the amount of time you spend watching your television, preferably to zero, then spend that time implementing other actions suggested in this chapter and in the list of suggested readings that follows the appendix.

There are lots of ways to have low-energy fun, especially with other people. Learn how to make things—our villages and towns once made all the things we used and wore in life. If you can play an acoustic, orchestral instrument, join or form an orchestra. If you can sing, join a choir. Dance to music played by live musicians using no electronic amplification whatsoever. Write, read, and perform plays. Tell stories. Observe nature, and tell others what you observe.

Understand how modern communications systems work, especially the Internet, which includes both e-mail and the Web. This will be an extremely important tool for sharing knowledge between communities. Form groups, organize.

Get political, stand for office, even though the system is monstrously broken and you think you have no chance of winning.

Practice "corporate disobedience": if you are not legally bound to buy something from a public-quoted corporation (that is a corporation that is listed on a stock exchange), then don't buy it. This type of disobedience is inspired by the idea of civil disobedience, but it has one major difference–corporate disobedience is entirely legal, by definition.

Money-Energy-Food Experiments

Many of the actions described here will depend upon differences in locale, and there are many others that could be added, but it is still quite a list. Nevertheless, these are not things that no one but a saint or millionaire could do. Some of these proposals are clearly going to be very difficult, or perhaps impossible in the short run, for many people, but some are possible for anyone. In the medium- to long-term, the only way to achieve this kind of change is with a new, low-energy public infrastructure, where cities, towns, and villages are once again designed for humans and not cars, and vital needs are met with locally produced energy, food, and money.

Trying "post-carbon" experiments with local money-energy-food systems is indispensable. Universities and high schools are ideal units, but anywhere with a strong sense of identity, cohesion, and place—and some soil, will be fertile ground for those willing to try. We will need a range of possibilities of local parallel public infrastructure so that as things get worse a gradual transfer can be made.

Don't Give Up

Some of the above measures don't look much like the usual green wish list. One of the things not strongly recommended is "efficiency" measures. The reason is that I think efficiency is a code word for business-as-usual lite, which is clearly doomed.[67] There are those who claim that a given technological process will save 50 percent or even more in energy usage, and that if we all used it, then the difference would be considerable. In theory, there may be some truth to this argument, but what is not said is that when we just use techno-fixes, we actually end up using more energy, because very often economic growth is stoked by the act of developing and buying the technology. Furthermore, the energy savings made often yield spare cash, which is frequently immediately spent on more energy-intensive devices, which more than soak up the energy or resource just saved. Thus ultimately more things are made and bought, fueling consumerism, euphoria, and confirming the idea that we can technologize our way out of problems.

The really good thing about using much less energy is that either way most of us will be better off. If a miracle happens, and the energy crisis somehow goes away tomorrow, then so be it, everyone's energy bills will be lower. If not, we have some chance of eating and staying warm.

I am not optimistic that the plea to use much less energy will be heeded by governments, and certainly nothing I have written here will be welcomed by the global corporations, whose extinction I am explicitly calling for, but I know for a fact that there are already people all over the world, and especially in North America, who want to be part of reversing the course we are on. We know that we must start sharing as if our lives depended on it—because they probably do.

For those who want to act, I hope this book will give you some useful facts, some good arguments, and a large dose of inspiration. We're going to need all we can get. And don't give up—we haven't got time. Millions of people want action, want to do something, want it to mean something, and want it to last. The time to act is now, but not alone, like Gary Cooper; rather we must work together, in small groups and large, and start replacing the future with something that makes sense on a planet we have abused for far too long.

Appendix: Key Numbers and Terms

Production: refers to gas (or oil) extracted from the earth. In this book I have also used the term "extraction," as a reminder that gas (and oil) are not "produced" in the way that barley or rice is. The industry always uses the word "production."

See chapter 2, for a discussion of reserve terminology (Proved, Probable, Possible, and so on).

The three industrialized countries who are likely to face a natural gas crisis soonest are the United States, Canada, and the British Isles (including the Irish Republic). New Zealand may not be far behind.

Table 1

| Country | | Consumption | | Production (Dry gas) | | Reserves |
		Per year (Tcf)	Per day (Bcf)	Per year (Tcf)	Per day (Bcf)	(Tcf)
US	(2003E)	22.3	61	19.5	53.4	183
Canada*	(2002E)	6.6	18	2.9	7.9	59.1 (2004E)
UK	(2002E)	3.3	9	3.6	9.9	22.2 (2004E)

See chapter 5 for an explanation of why Canada is also facing problems.

All the above are average amounts, since natural gas consumption varies greatly with season. Production also varies, though the factors for that are rather more complicated. The year of the data is given in parentheses, and most are estimates, marked with an *E*.

Amount of gas consumed by the United States in 1 year (2003):	22.3 Tcf
Fraction of world gas production consumed by United States:	24%
Fraction of world gas reserves held by United States:	1.4%
Months of highest gas consumption in United States (over 2 Tcf/month):	winter (Dec–Mar)

Other facts and figures

The following terms are used in the international natural gas industry.

Cubic feet and Btu are used to measure gas in the United States. The metric system prevails in many other places.

U.S. Terms

(Note: by law U.S. federal agencies are obliged to use SI (International System of) Units. However this is rarely done.)

1 cubic foot is approximately equal to 0.0283 cubic meters.

The other main term used to measure, sometimes on the same page of data, is Btu, British thermal unit. A Btu is a measure of energy, and its equivalence to a volume of natural gas will depend on composition of the gas. For the United States and much of the industrialized world (except Japan), 1 cubic foot of gas is taken as being equivalent to 1026 Btu. In energy terms, 1 Btu is equivalent to 1055 Joules or 1.055 kJ. A joule is the amount of work done when a force of one Newton is applied through a distance of one meter.

Tcf = trillion cubic feet.

1 Tcf = 1,000,000,000,000 cubic feet.

Bcf = billion cubic feet.

1 Bcf = 1,000,000,000 cubic feet.

MMcf = million (actually thousand thousand) cubic feet.

1 MMcf = 1,000,000 cubic feet.

Mcf = thousand cubic feet.

1 Mcf = 1,000 cubic feet. (Note: There is a real danger of confusion between the *M* in Mcf, which means thousand, *and m* in the metric system, which means mega or million.)

Btu = British thermal unit, which is a measure of energy. A Btu is the quantity of heat required to raise the temperature of 1 pound of water by 1 degree Fahrenheit.

MMBtu = Million Btu. 1 MMBtu is equivalent to about 975 cubic feet of gas (Note: MMBtu uses M to mean thousand, as in Mcf above. Thus, MMBtu, is thousand thousand Btu, or a million.)

1 Tcf is approximately equal to 1 Quad. 1 quad is 1 quadrillion Btus or $1x10^{15}$ Btus, equal to $293x10^9$ kWh. 1 Tcf is about 300 million MWh.

Metric

1 cubic meter = 35.315 cubic feet

1 Tcf = 28.3 billion cubic meters

The following terms are used in the metric system:

k = kilo or thousand (10^3 also written as 10^3)

M = mega or million (10^6 also written as 10^6)
(Note the possible confusion with Mcf, which means thousand cubic feet. See above.)

G = giga or billion (10^9 also written as 10^9)

T = tera or trillion   (10^{12} also written as 10^12)

P = peta or thousand trillion (10^{15} also written as 10^15)

E = exa or million trillion (10^{18} also written as 10^18)

World gas

World natural gas reserves – conventional (estimate)	10,000 Tcf
World natural gas reserves – unconventional (estimate)	2,500 Tcf
World natural gas produced in 2003:	92 Tcf

Roughly half of the world's discovered gas is in the fifty largest fields. The largest field, North Field-South Pars in Qatar-Iran, alone is reported to contain more than 10 percent. If so, then this field contains almost double the amount of energy found in Ghawar, Saudi Arabia, the world's largest oil field.

LNG

1 tonne of liquid natural gas is equal to about 1400 cubic meters of natural gas at normal temperature and pressure.

1 U.S. ton (2,000 lbs) =	0.907 tonne (U.S. tons are also known as short tons)
1 cubic meter =	35.315 cubic feet
1 U.S. ton LNG =	about 45,000 cubic feet or 45 Mcf in gaseous form
1 tonne LNG =	2.2 cubic meters in liquid form and about 50 Mcf in gaseous form
Largest LNG tanker (2004):	145,000 cubic meters, which expands to about 3 Bcf of gas (at normal temperature and pressure). Note: LNG tankers are normally quoted in metric measures.

Chapter 2, "The Gas Itself," discusses many aspects in depth. Production issues are covered in chapter 5.

Top 25 Gas producers in 2000

(Data from EIA)

Country	Production in Bcf
Russia	20631
United States	20002
Canada	7121
United Kingdom	3967
Algeria	3136
Netherlands	2559
Indonesia	2530
Iran	2299
Uzbekistan	1992
Norway	1914
Saudi Arabia	1864
Turkmenistan	1642
Malaysia	1521
United Arab Emirates	1474
Argentina	1472
Mexico	1315
Venezuela	1225
Qatar	1162
Australia	1155
China	957
Pakistan	856
India	821
Egypt	802
Germany	779
Thailand	713

Top 25 Gas reserves (in Tcf)

Data source:	Oil & Gas Journal	World Oil
Country	January 1 2003	Year-End 2002
Russia	1680	1700
Iran	812	914
Qatar	509	916
Saudi Arabia	225	231
United Arab Emirates	212	204
United States	183	189
Algeria	160	170
Venezuela	148	149
Nigeria	124	178
Iraq	110	113
Indonesia	92	73
Australia	90	85
Norway	77	75
Malaysia	75	88
Turkmenistan	71	NA
Uzbekistan	66	NA
Kazakhstan	65	NA
Netherlands	62	55
Canada	60	60
Egypt	58	59
China	53	47
Kuwait	53	53
Libya	46	46
Ukraine	40	NA
Azerbaijan	30	NA

Online Resources for Analysis and Action

The following is a short list of Web sites available at the time of writing that are designed to help the reader gain more familiarity with both the natural gas peak in North America and global oil peak, and some practical responses. As chapter 9 explains, given wider global circumstances, those facing gas peak cannot simply turn to oil (or anything else at suitable scale) to make up the difference. A general reduction in energy usage will therefore be necessary. Given the speed at which we shall have to do this, and combined with having reached the limits of so many other resources on the planet, it will become increasingly obvious that we must plan for contraction of the economy and the population. This will not be easy.

The oil and gas peak sites deal with the natural gas (and oil) supply problem primarily, but several of them do also confront demand reduction. The first four sites are those that I have founded or cofounded over the last few years to look at exploring, publicizing, and addressing the severe biosphere issues, particularly oil and gas peak. The rest of the sites are connected with analyzing and developing responses to resource constraints, and some cover all the categories mentioned below.

Sites Founded by Julian Darley

Global Public Media:
http://www.globalpublicmedia.com
Specializes in long-form interviews with experts and researchers in many areas to do with energy peak and decline, and also more generally with biosphere destruction.

Post Carbon Institute:
http://www.postcarbon.org
Think tank dedicated to developing large- and small-scale responses, both public and personal, to the decline of global oil, and natural gas decline in North America and other regions.

Citizens Committee on Oil Peak and Decline (COPAD):
http://www.copad.org
(cofounded with Jim Baldauf, Texas oil man)
Publicizing work on oil and gas peak specifically to increase public participation in policy making.

Community currencies:
http://www.communitycurrency.net
Information about local money systems, and development site of an online local money and citizens' directory.

Sites Addressing Oil and Gas Peak

ASPO, The Association For The Study Of Peak Oil & Gas:
http://www.peakoil.net
ASPO is a network of scientists, affiliated with European institutions
and universities, having an interest in determining the date and impact
of the peak and decline of the world's production of oil and gas, due to
resource constraints.

http://www.dieoff.org
A very large resource of core articles and papers about population
bloom on a planet with limited resources.

http://www.hubbertpeak.com
Seminal work on energy by Bartlett, Campbell, Cleveland, Deffeyes,
Duncan, Hubbert, Ivanhoe, Laherrère, Reynolds, Youngquist, and
much discussion of possible alternatives.

http://www.asponews.org
Unofficial archive of the ASPO Newsletter, a complete archive of all of
the newsletters and quick links to country assessments and country plots.

http://www.odac-info.org
The Oil Depletion Analysis Centre (ODAC) is an independent,
U.K.-registered educational charity working to raise international
public awareness and promote better understanding of the world's oil-
depletion problem.

http://www.simmonsco-intl.com/research.aspx?Type=msspeeches
Recent papers and speeches given by Matthew R. Simmons, energy
banker, Simmons and Company International.

http://www.museletter.com
Site of Richard Heinberg, author of "The Party's Over: Oil, War, and
The Fate of Industrial Societies," and "Powerdown." Heinberg dis-
cusses realities of contraction of population, economic, and energy.

http://www.oilcrash.com
A New Zealand perspective.

FEASTA (pronounced "fasta"):
http://www.feasta.org
An Irish perspective on oil and gas peak, renewable energy, local money,
and related work that is highly critical of economic growth.

Sites of Selected Government Energy Departments or Statistics

U.S. Energy Information Administration:
http://www.eia.doe.gov/
Official energy statistics from the U.S. government.

National Energy Board of Canada:
http://www.neb-one.gc.ca/index_e.htm

U.K. Energy:
http://www.dti.gov.uk/energy/

Sites about Local Money

(Also referred to as local currencies, community currencies, alternative currencies)

E. F. Schumacher Society:
http://www.schumachersociety.org
Pioneers in recent development and use of local money.

LETS:
http://www.gmlets.u-net.com
Information about the widespread local money system originally developed by Michael Linton. (LETS is sometimes known as Local Exchange Trading Systems, though LETS just means "let's do it").

new economics foundation (nef):
http://www.neweconomics.org
Founded in 1986, nef is based in Britain and works across a broad range of areas to revitalize local communities and economies.

Sites about Population Issues

Population Reference Bureau: http://www.popnet.org/
A range of official, academic, and NGO (nongovernmental organization) population-related websites.

Population Connection (Formerly called Zero Population Growth):
http://www.populationconnection.org
Aims to educate young people and advocates progressive action to stabilize world population at a level that can be sustained by Earth's resources.

Negative Population Growth (NPG):
http://www.npg.org
A membership organization, founded in 1972, working for a U.S. population commensurate with its resources.

Sustainable Population Australia (SPA):
http://www.population.org.au
Founded in 1988 to encourage informed public debate about how Australia and the world can achieve an ecologically sustainable population.

Sites about Renewable Energy

A Global Overview of Renewable Energy Resources (AGORES):
http://www.agores.org
European Union (EU) site, part of EU's attempt to reach 12 percent renewable energy sources (or RES) by 2010.

Home Power:
http://www.homepower.com
Hands-on journal of homemade power.

National Renewable Energy Laboratory:
http://www.nrel.gov
U.S. government-funded research into the technology of renewable energy. Pervaded by uncritical techno-optimism.

Sites about Food Production

The Occidental Arts and Ecology Center (OAEC):
http://www.oaec.org
Founded in 1994 by a group of biologists, horticulturists, educators, activists, and artists, OAEC has 80 acres of land on the California coast and comes very close to producing food with no oil or gas inputs. They also have courses and educational materials.

Ecology Action / Biointensive Mini-Farming:
http://www.growbiointensive.org & http://www.johnjeavons.info
Founded in 1972 by John Jeavons, Ecology Action has been researching, developing; and sharing traditional techniques for growing more food in a small area, using simple tools and seeds, while maintaining or increasing the health and productivity of the soil.

Permaculture:
http://www.permaculture.net
Information about permaculture, mainly in the United States.

Notes

Chapter 1: Introduction

1. Ontario also produces a large amount of power from its nuclear reactors, about which there is considerable controversy.
2. The words of John L. O'Sullivan (1845), quoted in Alan Brinkley, *American History,* A Survey, vol. 1, 9th ed. (New York: McGraw-Hill, 1995), quoted in Michael T. Lubragge, *Manifest Destiny: The Philosophy That Created a Nation,* http://odur.let.rug.nl/~usa/E/manifest/manif1.htm (accessed January 26, 2004).
3. Union of Concerned Scientists, "Clean Energy update—5/2004," http://www.ucsusa.org/clean_energy/renewable_energy/page.cfm?pageID=1404 (accessed May 24, 2004).
4. U.S.-Canada Power System Outage Task Force, "Interim Report: Causes of the August 14th Blackout in the United States and Canada," November 2003, http://reports.energy.gov/814BlackoutReport.pdf (accessed February 4, 2004).
5. Coalbed methane is explained in chapter 2.

Chapter 2: The Gas Itself

1. Jan B. van Helmont (1580–1644). Energy Underground, "The Archive," http://cinergy.energyunderground.com/hiband/zone/history/helmont.html (accessed January 27, 2004).
2. Coalbed methane is an example of slow extraction, but low extraction costs, at least in dollar terms. On the other hand, deepwater and polar gas are often extracted as fast as possible for economic reasons but invariably require very expensive equipment.

3. National Petroleum Council, "Balancing Natural Gas Policy: Fueling the Demands of a Growing Economy," September 2003, http://www.npc.org /reports/ng.html (accessed February 10, 2004).

4. Or about 2.3 million cubic meters.

5. This section owes much to background briefings at NaturalGas.org, http://www.naturalgas.org/overview/background.asp (accessed February 4, 2004); and "What Is Crude Oil?" Chevron, http://www.chevron.com/learning _center/crude/ (accessed February 4, 2004); and personal correspondence with Jean Laherrère, Colin Campbell, and Walter Youngquist.

6. "What Is Vitrinite?" http://www.ucl.ac.uk/~ucfbrxs/Vitrinite/VRnotes.pdf (accessed February 13, 2004). Inertinite is also a form of kerogen, but as the name suggests, it produces neither oil nor gas.

7. Colin Campbell, personal communication.

8. The sinking of large portions of Earth's crust.

9. A "geothermal gradient" is "the rate of increase in temperature per unit depth in the earth. Although the geothermal gradient varies from place to place, it averages 15°F/1000 ft or 25 to 30°C/km." From "Oilfield Glossary," Schlumberger Limited, http://www.glossary.oilfield.slb.com/Display.cfm?Term=geothermal %20gradient (accessed January 27, 2004).

10. A "hydrocarbon kitchen" is "an area of the subsurface where source rock has reached appropriate conditions of pressure and temperature to generate hydrocarbons; also known as source kitchen, oil kitchen or gas kitchen." "Oilfield Glossary," Schlumberger Limited, http://www.glossary.oilfield.slb.com/Display .cfm?Term=hydrocarbon%20kitchen (accessed January 27, 2004).

11. Jean Laherrère, quoting Ray Leonard of Yukos, private correspondence.

12. Most oil is from two brief epochs of intense global warming 90 and 145 million years ago in the mid-Cretaceous and late Jurassic. Much of what was formed earlier was lost.

13. Gas fields in the Po Valley of Italy produce biogenic gas, being continually recharged from shallow source rocks.

14. "The Formation of Natural Gas," http://www.naturalgas.org/overview/background .asp (accessed February 9, 2004).

15. The process has been replicated in the laboratory, but it is not the source of commercial gas known to the industry.

16. Only evaporites are complete seals. Evaporites are "a class of sedimentary minerals and sedimentary rocks that form by precipitation from evaporating aqueous fluid. Common evaporite minerals are halite, gypsum and anhydrite, which can form as seawater evaporates, and the rocks limestone and dolostone." "Oilfield Glossary," Schlumberger Limited, http://www.glossary.oilfield.slb.com /Search.cfm (accessed January 27, 2004).

17. But they are not filled nearly fast enough to make up for industrial-scale extraction!

18. Calcium sulfate, $CaSO_4$.
19. Except for extra-heavy oils.
20. Large finds are what the world depends on for oil. In 2003, for the first time in a century, there were no finds over 500 million barrels. Eighty percent of the world's oil comes from such deposits.
21. Deep water is a term that has no agreed definition but is used in the industry. In the Gulf of Mexico it means more than 600 feet (200 meters). Very deep water means drilling in thousands of feet of water. As technology improves, ever greater depths become possible, which is at least one reason why the meaning of deep is not fixed.
22. Exodus 3:2: "And the angel of the Lord appeared unto him in a flame of fire out of the midst of a bush: and he looked, and, behold, the bush burned with fire, and the bush was not consumed."
23. John R. Hale, Jelle Zeilinga de Boer, Jeffrey P. Chanton, and Henry A. Spiller, "Questioning the Delphic Oracle," *Scientific American* (August 2003), http://www.sciam.com/article.cfm?articleID=0009BD34-398C-1F0A-97AE 80A84189EEDF (accessed May 25, 2004).
24. To help people detect leaking gas, foul smelling chemicals called mercaptans are added to natural gas in tiny quantities.
25. The first factory was gaslit in 1782 in Britain. Street lighting soon followed. The year 1810 saw the first gaslighting in the United States, at Newport, Rhode Island. By 1823, over fifty towns and cities in Britain were lit by gas.
26. In the United States, the first commercial oil well was drilled in 1859 in Titusville, Pennsylvania.
27. Thirty feet is 9.14 meters.
28. The "derrick" is the tall metal pyramid used to support and lift the drill string.
29. About 3,000 meters.
30. Approximately 10 centimeters.
31. About 70 metric tonnes.
32. The azimuth is "the angle between the vertical projection of a line of interest onto a horizontal surface and true north or magnetic north measured in a horizontal plane, typically measured clockwise from north." In other words, it is the compass direction of the horizontal part of the hole. "Oilfield Glossary," Schlumberger Limited, http://www.glossary.oilfield.slb.com/Display.cfm?Term =azimuth (accessed January 29, 2004).
33. 30 to 45 centimeters.
34. With an onshore well, if oil is also present, as is the case with associated gas, the oil goes into a battery of crude oil tanks that are emptied into tanker trucks that haul crude from the oil fields to gathering locations or straight to refineries.
35. "Open hole" or "barefoot completion" simply means that no concrete or other tubing is used in a production well, and that the oil and gas flow naturally through the holes drilled in the surrounding rock.

36. My thanks to Jim Baldauf, a classic Texan oilman, of the best sort, for his help in this description.

37. About a meter or two-thirds of a meter.

38. Permeability is "the ability, or measurement of a rock's ability, to transmit fluids, typically measured in darcies or millidarcies." "Oilfield Glossary," Schlumberger Limited, http://www.glossary.oilfield.slb.com/Display.cfm?Term=permeability (accessed February 9, 2004).

39. From the U.S. Energy Information Administration, "Estimation of Reserves and Resources, Glossary G-4 & G," http://www.eia.doe.gov/pub/oil_gas/natural _gas/data_publications/crude_oil_natural_gas_reserves/current/pdf/appg.pdf (accessed February 9, 2004). The appendix G continues: "Reservoirs are considered proved if economic producibility is supported by actual production or conclusive formation test (drill stem or wire line), or if economic producibility is supported by core analyses and/or electric or other log interpretations."

40. This applies more to oil than gas, because, as noted above, gas has a much higher recovery factor than oil, often as high as 80 percent.

41. United States Geological Survey.

42. Energy Information Administration, "U.S. Crude Oil, Natural Gas, and Natural Gas Liquids Reserves 2002 Annual Report. Appendix G, Estimation of Reserves and Resources," http://www.eia.doe.gov/pub/oil_gas/natural_gas/data_publications/crude_oil_natural_gas_reserves/current/pdf/appg.pdf (accessed May 24, 2004).

43. Schlumberger defines 2P thus: "Reserves where commercial productivity has not been demonstrated, and there is a 50% probability that at least the sum of the estimated proved plus probable reserves will be recovered." http://www.sis. slb.com/media/about/whitepaper_oilgasreserve.pdf (accessed February 9, 2004).

44. "Oil and Gas Reserves," Schlumberger Limited, http://www.sis.slb.com /media/about/whitepaper_oilgasreserve.pdf (accessed February 9, 2004).

45. The exemption is discretionary from the Alberta Securities Commission and in fact depends on being quoted on both Canadian and U.S. stock exchanges. Only the largest companies fit this criterion. See http://www.albertasecurities. com/index.php?currentPage=98 (accessed March 26, 2004) for a list of the companies currently granted exemption. See http://www.albertasecurities.com/dms /1144/2232/5693__1211626_v7_-_51-101CP_&_APPENDICES.pdf (accessed March 26, 2004), Section 8, for an explanation of exemption, and see http://www.albertasecurities.com/dms/6247/6248/9553_51-102_factsheet.pdf (accessed March 21, 2004) for further explanation of the new 51-101 ruling, and http://www.albertasecurities.com/docHistory.php?project=2232&pageID=222 (accessed March 26, 2004) for a listing of all documents associated with the ruling. The ruling is significant, both politically and financially, because of the importance of perceived reserves.

46. http://www.sis.slb.com/media/about/whitepaper_oilgasreserve.pdf (accessed February 10, 2004).

47. "Oil and Gas Reserves Classification," Wood Mackenzie, private document; and Jean Laherrère, in personal communication.

48. In the vast but declining gas fields of western Siberia, it is said that C2 reserves are now being produced.

49. In 2004, there is heated, and increasingly open, debate about whether the world is at the peak of oil production.

50. Some have called for an international body to draw up an inventory of remaining world oil and gas. This may only be done after the reserves have been emptied.

51. The same kinds of reserve reporting and definitional problems also very much apply to oil.

52. U.S. Securities and Exchange Commission, "Topic 12: Oil and Gas Producing Activities," http://www.sec.gov/interps/account/sabcodet12.htm.

53. Permeability is "the ability, or measurement of a rock's ability, to transmit fluids, typically measured in darcies or millidarcies." "Oilfield Glossary," Schlumberger Limited, http://www.glossary.oilfield.slb.com/Display.cfm?Term=permeability> (accessed February 9, 2004).

54. Ted McCallister, "Impact of Unconventional Gas Technology in the Annual Energy Outlook 2000," U.S. Energy Information Administration, http://www.eia.doe.gov/oiaf/analysispaper/uncon_fig1.html (accessed February 10, 2004). No publishing date given except this: "Page last modified on Tue Jul 30 2002."

55. "Annual Energy Outlook 2004 with Projections to 2025," U.S. Energy Information Administration, http://www.eia.doe.gov/oiaf/aeo/gas.html (accessed February 10, 2004).

56. If the two are considered as separate items, which they are in the "Annual Energy Outlook 2004," U.S. Energy Information Administration.

57. EIA 1999: "Natural gas production 1990-2020," figure 1, http://www.eia.doe.gov/oiaf/analysispaper/uncon_fig1.html (accessed February 10, 2004).
EIA 2003: "Natural Gas Production by Source, 1990–2025," figure 87, in "Annual Energy Outlook 2004 with Projections to 2025," U.S. Energy Information Administration, http://www.eia.doe.gov/oiaf/aeo/gas.html (accessed February 10, 2004).

58. Several of the largest of which were already discovered forty years ago.

59. Over 10,000 wells as of 2000. See Mary Griffiths and Chris Severson-Baker, "Unconventional Gas: The Environmental Challenges of Coalbed Methane Development in Alberta," Pembina Institute publication June 2003.

60. "Coalbed Methane Basics," U.S. Energy Information Administration, http://www.eia.doe.gov/emeu/finance/sptopics/majors/coalbox.html (accessed February 4, 2004).

61. Griffiths and Severson-Baker, "Unconventional Gas: The Environmental Challenges of Coalbed Methane Development in Alberta."

62. Kristin Keith, Jim Bauder, and John Wheaton, "Coal Bed Methane: Frequently Asked Questions," Montana State University, Bozeman, Montana Bureau of Mines and Geology, http://waterquality.montana.edu/docs/methane/cbmfaq .shtml (accessed February 4, 2004).

63. Dick Wolfe and Glenn Graham, "Water Rights and Beneficial Use of Coal Bed Methane Produced Water in Colorado by October 2002," http://water.state.co .us/pubs/Rule_reg/coalbedmethane.pdf (accessed January 29, 2004).

64. U.S. coal reserves are reported as 275 billion short tons (12/31/98). "United States of America," U.S. Energy Information Administration, http://www.eia.doe.gov/emeu/cabs/usa.html (accessed February 4, 2004).

65. Griffiths and Severson-Baker, "Unconventional Gas: The Environmental Challenges of Coalbed Methane Development in Alberta."

66. Canada has 7.2 billion short tons (2001). "Canada," U.S. Energy Information Administration, http://www.eia.doe.gov/emeu/cabs/canada.html (accessed February 4, 2004).

67. "U.S. Crude Oil, Natural Gas, and Natural Gas Liquids Reserves, 2002 Annual Report," U.S. Energy Information Administration, p. 12, http://www.eia .doe.gov/pub/oil_gas/natural_gas/data_publications/advanced_summary _2002/adsum2002.pdf (accessed February 4, 2004).

68. Very little water has to be drawn off, and gas production starts relatively quickly.

69. Donovan Webster and Michael Scherer, "No Clear Skies," *Mother Jones* (October 2003), http://www.motherjones.com/news/feature/2003/09/ma_496_01.html (accessed February 4, 2004).

70. Explanation of salinity and sodium hazard in "Why Are People Concerned about CBM Product Water?" http://waterquality.montana.edu/docs/methane/cbmfaq .shtml (accessed February 4, 2004).

71. Keith, Bauder, and Wheaton, "Coal Bed Methane: Frequently Asked Questions."

72. Powder River Basin Resource Council, http://www.powderriverbasin.org (accessed February 4, 2004).

73. About 100,000 liters.

74. Just over 30 hectares.

75. 190,000 liters.

76. Stephen A. Holditch, "The Increasing Role of Unconventional Reservoirs in the Future of the Oil and Gas Business," Schlumberger Limited, December 3, 2001, http://www.spegcs.org/attachments/studygroups/6/Holditch2001.pdf (accessed February 4, 2004).

77. Jean H. Laherrère, "Hydrates: Some Questions from an Independent Oil and Gas Explorer," 2002, Introduction as chairman of RFP 9 "Economic Use of Hydrates: Dream or Reality?" WPC Rio de Janeiro, September 5, 2002, http://www.oilcrisis.com/laherrere/hydratesRio/ (accessed February 13, 2004).

78. Arthur H. Johnson, "Gas Hydrate Exploration in the Gulf of Mexico," American Chemistry Society, March 2003, http://oasys2.confex.com/acs/225nm /techprogram/P621998.HTM (accessed February 4, 2004); Lawrence M. Cathles III and Steven Losh, "Massive Hydrocarbon Venting with Minor, Constantly Replenished (Flow-through) Retention in a 100 x 200 km Area Offshore Louisiana Gulf of Mexico," American Chemistry Society, March 2003, http://oasys2.confex.com/acs/225nm/techprogram/P623430.HTM (accessed February 4, 2004).

79. "Occurrences of Natural Methane Hydrate," National Energy Technology Laboratory, January 26, 2004, http://www.netl.doe.gov/scng/hydrate/about -hydrates/occurrences.htm (accessed January 30, 2004).

80. http://www.netl.doe.gov/scng/hydrate/about-hydrates/safety-stability.htm (accessed January 30, 2004).

81. Neoliberalism is an extreme form of economic fundamentalism and capitalism, which emphasizes the power of free markets, free trade, total selfishness and uncontrolled greed, and seeks to remove all remaining constraints on corporations to do as they please. Neoliberals advocate the privatization of everything that can possibly make a profit for corporate shareholders. Neoliberalism is thus really a synonym for globalization. In 2004, neoliberal governments hold power in the United States, Canada, Britain, and Australia. Recently, the New Zealand government appears to be moving away from neoliberalism. South Africa shows strong signs of neoliberalism.

82. Or hope that fusion, or any other "bottomless" cheap energy scheme will work.

83. Wikipedia, Benito Mussolini, http://en.wikipedia.org/wiki/Benito_Mussolini #Quotes (accessed May 24, 2004).

84. "Methane Digesters," http://dir.yahoo.com/Science/Energy/Methane_Digester/ (accessed February 4, 2004); Allen Dusault, "Methane Digesters for Dairies: New Opportunities for Industry and the Environment," Ecological Farming Association, http://www.ecofarm.org/sa/sa_dairy_synopsis_digester.html (accessed February 4, 2004).

85. The process will operate even more effectively at 130°F, but it is much more sensitive to temperature changes.

86. It is a multistep process using different organisms for each part. In the first step, the organic substances are divided into molecular components (sugar, amino acids, glycerine, and fatty acids). Microorganisms convert these intermediate products into hydrogen and carbon dioxide, which are then transformed into methane and water according to the equation: $CO_2 + 4H_2 \longrightarrow CH_4 + 2H_2O$. "Biogas" Bioenergie, http://www.fnr-server.de/cms35/index.php?id=399 (accessed February 4, 2004).

87. "Methane (Biogas) from Anaerobic Digesters," Office of Energy Efficiency and Renewable Energy, January 2003, http://www.eere.energy.gov/consumerinfo /refbriefs/ab5.html, (accessed February 4, 2004).

88. Associated Press, "Utility Converts Cow Manure into Energy," December 6, 2001, http://www.climateark.org/articles/2001/4th/utcocowm.htm, (accessed February 4, 2004).

89. AgStar Digest, Environmental Protection Agency, Winter 2003, http://www.epa.gov/agstar/library/2002digest.pdf, (accessed February 4, 2004).

90. "Generating Methane Gas From Manure," http://muextension.missouri.edu/explore/agguides/agengin/g01881.htm (accessed March 21, 2004).

91. Mike Ewall, "Primer on Landfill Gas as 'Green' Energy," Pennsylvania Environmental Network, February 10, 2000, http://www.penweb.org/issues/energy/green4.html, (accessed February 4, 2004).

Chapter 3: Moving Gas

1. There are three regional markets: North America, Europe, and Asia Pacific.

2. http://www.platts.com/features/usgasguide/pipelinemap.shtml (accessed July 24, 2003).

3. http://www.phillips66.com/energyanswers/naturalgas.htm (accessed May 25, 2003).

4. Jerry White, "Who Is Responsible for the Oil Explosion in Nigeria?" World Socialist, October 21, 1998, http://www.wsws.org/news/1998/oct1998/nig-o21.shtml (accessed February 3, 2004).

5. The Institute for the Analysis of Global Security's "Iraq Pipeline Watch" lists thirty-seven pipeline or related attacks from June 2003 to the end of January 2004. http://www.iags.org/n013104.htm (accessed February 12, 2004).

6. Shirin Wheeler, "EU Probes Burma Pipeline Abuses," BBC News, October 12, 2001, http://news.bbc.co.uk/1/hi/world/europe/1596181.stm (accessed February 12, 2004).

7. For example, EarthRights International, http://www.earthrights.org/; Amazon Watch, http://www.amazonwatch.org/; Tourism Watch, http://www.tourism-watch.de; CorpWatch, http://www.corpwatch.org/ (all accessed February 12, 2004).

8. "Corporates," Friends of the Earth, July 2002, http://www.foe.co.uk/campaigns/corporates/press_for_change/email_westlb/ (accessed February 3, 2004).

9. U.S. Department of Energy, Office of Fossil Energy, "Natural Gas Import & Export Regulation-Authorizations," updated on 08/21/2003, http://www.fe.doe.gov/programs/gasregulation/authorizations/a_questions.html (accessed May 24, 2004).

10. The Phillips case was: *Phillips Petroleum Co. v. Wisconsin Public Service Commission.* See Baker Communications Inc., "From the Wellhead to Burner Tip: Basics of Natural Gas Online Training," http://www.bakercommunications.com/whbt/index.htm (accessed February 3, 2004). The ruling undoubtedly

kept prices low, which eventually encouraged too much demand without stimulating enough supply. In hindsight, a low price ironically slowed down the depletion of U.S. gas fields by discouraging producers.

11. Since much gas was discovered whilst looking for oil, which has been a global product almost from its commercial inception, it is likely that the dampener referred to was more upon gas production than exploration.

12. It is interesting to note that the U.S. gas crisis followed the peak of gas production by about three years, just as the oil crisis of 1973 came three years after the U.S. oil peak.

13. "History," NaturalGas.org, http://www.naturalgas.org/overview/history.asp (accessed February 3, 2004).

14. U.S. Federal Energy Regulatory Commission Order 436

15. U.S. Federal Energy Regulatory Commission Orders 636 and 637

16. Except for the relatively small offshore gas plays of eastern Canada, though even these may not grow much more. A play may be described as oil or gas production from reservoirs of similar geological characteristics within a field.

17. Some petroleum geologists put the U.S. oil production peak in 1971. It is partly a matter of definitions. No one disagrees that it was around this time.

18. In July 2003, the Canadian population was estimated at 31,629,677. M. David Bennett, "Teaching and Learning about Canada," January 6, 2004, http://www.canadainfolink.ca/canfo.htm (accessed January 30, 2004).

19. Canadian conventional oil production peaked in 1973 but then maintained a lower, gently rising plateau until 1999. Colin Campbell, "Production by Country and Peak Date," Association for the Study of Peak Oil and Gas, http://www.asponews.org/docs/All-Peak.xls (accessed January 30, 2004).

20. Richard F. Meyer and Emil D. Attanasi, "Heavy Oil and Natural Bitumen—Strategic Petroleum Resources," U.S. Geological Survey, Fact Sheet 70-03, August 2003, http://pubs.usgs.gov/fs/fs070-03/fs070-03.html (accessed February 12, 2004).

21. "Oil equivalent," in this case, means that though there are lots of hydrocarbon molecules, they aren't in an immediately useable form.

22. Marianne Lavelle, "New Technology Makes It Easier to Tap Canada's Oil Reserves," Petroleumworld.com, October 8, 2003, http://www.petroleumworld.com/story2545.htm. (accessed February 3, 2004).

23. Depending on the method of extraction and processing, tar sands production requires between 600 cubic feet and 1,400 cubic feet of gas per barrel of synthetic crude oil that is sent to market.

24. "Canada's Energy Future: Scenarios for Supply and Demand to 2025," http://www.neb.gc.ca/energy/SupplyDemand/2003/English/SupplyDemand2003_e.pdf (accessed February 12, 2004).

25. Bcf means billion cubic feet.

26. The state of Canada's gas reserves and production capacity will be covered in chapter 5.

27. Jeffrey Jones, "Canada Oil Output Lags Forecasts, '04 Seen Muted," Reuters, November 14, 2003, http://www.forbes.com/markets/newswire/2003/11/14 /rtr1148906.html (accessed February 12, 2004).

28. "The first liquefied natural gas plant was built in West Virginia in 1912, while the first commercial liquefication plant was built in Cleveland, Ohio, in 1941." History of Liquefied Natural Gas, http://www.dom.com/about/gas-transmission /covepoint/lng_history.jsp (accessed March 23, 2004).

29. Wood Mackenzie, personal communication.

30. This boil-off gas supplies about half of the ship's fuel, the other half is from traditional bunker oil, the nautical name for the diesel carried on board a ship to power its engines.

31. A knot is 1 nautical mile per hour. A nautical mile is 6,076 feet, whereas statute mile is 5,280 feet. A nautical mile is thus approximately 15 percent longer.

32. Typically, 0.15 to 0.25 percent of the cargo is consumed per day, during which the tanker will travel about 480 nautical miles (about 550 miles or 890 km). "Mid-Term Natural Gas Supply: Analysis of LNG Imports," http://www.eia .doe.gov/oiaf/servicerpt/natgas/chapter3.html (accessed February 12, 2004).

33. Main data from Wood Mackenzie, personal communication.

34. World Energy Outlook, "The World Energy Investment Outlook 2004," November 4, 2003, International Energy Agency, http://www.worldenergyoutlook .org/weo/pubs/gio2003.asp (accessed February 3, 2004).

35. CH·IV International, "A Brief History of U.S. LNG Incidents," http://www.ch-iv.com/lng/incid1.htm (accessed February 3, 2004).

36. Ibid.

37. "Kellogg Brown & Root Passes All Performance Tests on New LNG Revamp Project in Algeria," press release 1999, Halliburton Company, http://www.halliburton .com/news/archive/1999/kbrnws_102899.jsp (accessed January 31, 2004).

38. "Industry: Skikda Accident Would Not Happen in the U.S.," *Oil & Gas Journal,* January 27, 2004, http://ogj.pennnet.com/articles/web_article_display.cfm ?Section=OnlineArticles&ARTICLE_CATEGORY=GenIn&ARTICLE_ID=19 7389 (accessed February 3, 2004).

39. Simon Romero, "Algerian Explosion Stirs Foes of U.S. Gas Projects," *New York Times,* February 12, 2004, http://www.nytimes.com/2004/02/12/business /worldbusiness/12gas.html (accessed February 12, 2004).

40. "Sabotage on Gas Line between Hassi R'mel and Arzew," Africa Intelligence, http://www.africaintelligence.com/ps/AN/Arch/AEM/AEM_166.asp (accessed January 30, 2004).

41. "Algeria," Country Analysis Briefs, U.S. Energy Information Administration, January 2003, http://www.eia.doe.gov/emeu/cabs/algeria.html (accessed January 31, 2004).

42. Ben Raines and Bill Finch, "Report Plays Down Threat," *The Sacramento Bee,* October 26, 2003.

43. Ibid.

44. Ibid.

45. Ibid.

46. Ibid.

47. Global oil peak may also have triggered a world oil crisis by then.

48. Matt Simmons, http://www.simmonsco-intl.com (ongoing website).

49. Wood Mackenzie, personal communication.

50. France is now also building LNG tankers.

51. Gregson Vaux, LNG tanker expert. Posted July 9, 2003, to Energy Resources list, http://groups.yahoo.com/group/energyresources/

52. Wood Mackenzie, personal communication.

Chapter 4: Demanding Gas

1. The concealment from the public of global oil peak was beginning to crack in 2003. For instance, *CNN* reported the work of the Association for the Study of Peak Oil and Gas (ASPO); the *Guardian* published George Monbiot's clear exposition of oil peak; the *BBC Money Programme* produced a documentary linking the 2003 Iraq invasion to oil peak.

2. $E=mC^2$ (E=energy, m=mass, C=speed of light).

3. The four laws of thermodynamics: 0) If two systems are in thermal equilibrium with a third system, then they must be in thermal equilibrium with each other [this law is sometimes called the Zeroth Law, because it was added later]; 1) Conservation of Energy: energy cannot be created or destroyed, only transferred from one system to another; 2) Entropy: all natural processes are irreversible and disorder increases over time as systems move towards equilibrium [the famous Second Law]; 3) Absolute Zero is the lowest temperature, when complete disorder is achieved in the universe and there is no entropy. These laws can be memorably combined into: $dE - TdS + PdV \leq 0$ where energy is E, temperature T, pressure P, and volume V. [This law is the answer to why Zero-Point Energy won't work.] http://scienceworld.wolfram.com/physics/CombinedLawofThermodynamics.html (accessed February 13, 2004).

4. In the early months of 2004 North American papers of record finally became aware of oil peak, and particularly that Saudi Arabia, containing the world's largest and most important oil field, may not be able maintain output for much longer.

5. The six main greenhouse gases referred to in the Kyoto Protocol are carbon dioxide (CO_2), methane (CH_4), nitrous oxide (N_2O), hydrofluorocarbons (HFCs), perfluorocarbons (PFCs), sulfur hexafluoride (SF_6).

6. "Kyoto and Beyond," Friends of the Earth, 1997, http://www.foe.co.uk/resource /briefings/kyoto_beyond.html (accessed February 1, 2004).

7. "Bonn Compromise Saves Kyoto Process," D+C Development and Cooperation, no. 6, (November/December 2001), p. 29, http://www.dse.de/zeitschr/de601-12.htm (accessed February 1, 2004).

8. By early 2004 implementing the protocol would mean a 30 percent cut in emissions for Canada.

9. "No Need for Russia So Far to Ratify Kyoto Protocol-Statesman," *Itar-Tass,* January 31, 2004, http://www.itar-tass.com/eng/level2.html?NewsID=370096 &PageNum=0 (accessed February 1, 2004).

10. Reuters, "EU Links Russia's WTO Entry to Kyoto," January 29, 2004, http://www.planetark.com/avantgo/dailynewsstory.cfm?newsid=23616 (accessed February 1, 2004).

11. The United Kingdom converted to gas for economic reasons, as mentioned, and German unification reduced the use of East German lignite, a particularly dirty form of coal.

12. "U.K. Elections Background," *BBC,* February 28, 1974, http://www.bbc.co.uk /election97/background/pastelec/ge74feb.htm (accessed February 11, 2004).

13. "U.K. Government Confesses to Peak Oil and Gas," ASPO Newsletter no. 36 (December 2003), http://www.asponews.org/HTML/Newsletter36.html (accessed February 1, 2004).
"National Grid Transco Agrees 20-Year Capacity Contract for Isle of Grain LNG Terminal," National Grid Transco, October 24, 2003, http://www.ngtgroup .com/media/press/mn_pr_24102003.html (accessed February 1, 2004).

14. Severin Carrell, "Ban Fishing in Third of All Seas, Scientists Say," *Independent/UK,* August 31, 2003, http://www.commondreams.org/headlines03 /0831-05.html (accessed February 11, 2004).

15. "The Promise of a Blue Revolution," *The Economist,* August 7, 2003, http://www.economist.com/business/displayStory.cfm?story_id=1974103 (accessed February 1, 2004). The problems with fish farming have also become apparent: "Call for Salmon Safety Audit Scottish Fish Farms Came out Bottom in U.S. Study," http://news.bbc.co.uk/1/hi/scotland/3403237.stm (accessed February 1, 2004); Stephen Khan, "Tainted Salmon: Farming Methods Turning Health Food into Poison," *The Observer,* January 17, 2004, p. 9, http://www.taipeitimes.com/News/edit/archives/2004/01/17/2003091729 (accessed February 1, 2004).

16. The key laws of thermodynamics were not articulated until the nineteenth century.

17. Sometimes referred to as EROEI—energy returned on energy invested.

18. Ghawar is estimated to contain about 120 billion barrels of oil.

19. Though many agree that this decade will be a petroleum plateau. In 2004 the evidence was mounting fast that the world was at or very close to the oil peak-plateau.

20. "Now That We've Found It, What Do We Do With It?" Energy Underground, http://cinergy.energyunderground.com/hiband/zone/science/what.html (accessed February 1, 2004).

21. "Balancing Natural Gas Policy," National Petroleum Council, 2003, http://npc.org/NG_Volume_1.pdf (accessed February 12, 2004).

22. "International Energy Outlook 2003," U.S. Energy Information Administration, May 2003, http://www.eia.doe.gov/oiaf/ieo/ (accessed February 12, 2004).

23. In 2003, gas supplied almost as much of the extra power needed for summer as did coal, such that while coal continued to supply about half of U.S. electricity, gas leapt from 15 percent in April (traditionally the lowest month of demand) to over 21 percent in August (often the peak month). The amount of gas used in August was almost double that in April. Data from the U.S.Energy Information Administration in "Net Generation by Energy Source: Total (All Sectors), 1990 through September 2003," http://www.eia.doe.gov/cneaf/electricity/epm/table1_1.html (accessed February 1, 2004).

24. Andrew Weissman, "Natural Gas Supply, Demand and Pricing: Puncturing Natural Gas Myths," part 1, November 21, 2003, http://www.energypulse.net/centers/article/article_display.cfm?a_id=556 (accessed February 11, 2004).

25. Even in December 2003, with figures that show clearly that 2002 U.S. gas production had fallen, the "Annual Energy Outlook 2004 with Projections to 2025" still predicts that the U.S. gas supply will grow. http://www.eia.doe.gov/oiaf/aeo/gas.html (accessed February 12, 2004).

26. "Is Britain Running on Empty?" July 17, 2002, Alexander's Oil and Gas Connection, http://www.gasandoil.com/goc/news/nte23465.htm (accessed February 1, 2004).

27. "BP Forms Gas Joint Venture with Sonatrach," October 31, 2003, Alexander's Gas & Oil Connection, http://www.gasandoil.com/goc/company/cna34842.htm (accessed February 1, 2004).

28. Alexandra Fouché, "France's Nuclear Option," *BBC News Online,* October 9, 2003, http://news.bbc.co.uk/1/hi/world/europe/3177360.stm (accessed February 1, 2004); "The French Fast-Breeder Programme," http://www.wise-paris.org/english/ournewsletter/1/page6.html (accessed February 1, 2004); Yves Marignac, "L'EPR : un choix du passé qui fermerait l'avenir?" November 2003, http://www.wise-paris.org/francais/rapports/notes/031117NoteEPR-Passe.pdf (accessed February 1, 2004).

29. Technically called "hydronic" heating.

30. Vaclav Smil, *Enriching the Earth: Fritz Haber, Carl Bosch, and the Transformation of World Food Production* (Cambridge, Mass.: MIT Press, 2001).

31. Vaclav Smil, "Detonator of the population explosion," Nature 400 (July 29, 2003): 415. Reproduced at http://home.cc.umanitoba.ca/~vsmil/pdf_pubs /nature7.pdf (accessed March 23, 2004).

32. "Natural Gas: Domestic Nitrogen Fertilizer Production Depends on Natural Gas Availability and Prices," United States General Accounting Office, September 2003, http://www.gao.gov/new.items/d031148.pdf (accessed February 1, 2004).

33. "Natural Gas Inventory," U.S. Energy Information Administration, January 30, 2004, http://tonto.eia.doe.gov/dnav/ng/ng_pri_sum_nus_m_d.htm (accessed February 1, 2004).

34. Letter to NPC Summit from Gregori Lebedev, president and chief executive officer, American Chemistry Council, June 25, 2003, http://accnewsmedia .com/docs/1300/1234.pdf (accessed February 12, 2004).

35. The term "Green Revolution" was coined in the 1960s to describe the development of "improved" seed varieties that were more responsive to controlled irrigation and petrochemical fertilizers, thus increasing the conversion efficiency of industrial inputs, largely derived from oil and gas, into food. See Peter Rosset, Joseph Collins, and Frances Moore Lappé, "Lessons from the Green Revolution," Tikkun (March/April 2000), http://www.foodfirst.org/media/opeds/2000/4-greenrev.html (accessed February 1, 2004).

36. Each pound of grain requires 400 to 500 pounds of water, depending on type and conditions.

37. "Agriculture Subcommittee Opens Comprehensive Farm Fuel Hearings," U.S. House Committee on Agriculture, April 25, 2001, http://agriculture.house .gov/107/pr010425e.html (accessed February 12, 2004).

38. Laura Skillman, "Higher Fuel Prices Mean Higher Grain Drying Costs," University of Kentucky, September 17, 2003, http://www.ca.uky.edu/agc /news/2003/Sep/drying.htm (accessed February 12, 2004).

39. "From Natural Gas to Plastic Cups," MG Engineering, September 11, 2003, http://213.156.165.2/english/nbsp/menu/media/news/polypropylene_for_the_f irst_time_produced_from_natural_gas_using_the_mtpr_process/ (accessed February 12, 2004).

40. "Thermal Plasma Conversions of Methane to Acetylene and Hydrogen," Idaho Engineering and Environmental Laboratory, http://energy.inel.gov/fossil/refine /thermal.shtml (accessed February 12, 2004).

41. "Petrochemicals and Natural Gas Prices: Short-Term Pain, Long-Term Concern," Federal Reserve Bank of Dallas, January 2001, http://www.dallasfed .org/research/houston/2001/hb0101.pdf (accessed February 12, 2004).

42. "Kansas Rocks, Minerals, and Petroleum Resources," Kansas Geological Survey, http://www.kgs.ukans.edu/Extension/KGSrocks/earthresources.html (accessed February 12, 2004).

43. "Why Natural Gas Vehicles?" U.S. Department of Energy, May 16, 2003, http://www.ott.doe.gov/ngvtf/why.html (accessed February 12, 2004).

44. "Urban Air: Health Effects of Particulates, Sulfur Dioxide, and Ozone," World Resources Institute, 1998–99, http://www.wri.org/wr-98-99/urbanair.htm (accessed February 12, 2004).

45. "Methane and Other Gases," U.S. Environmental Protection Agency, July 2003, http://www.epa.gov/ghginfo/ (accessed February 12, 2004).

46. "Natural Gas Buses—No Better Than Diesel Buses?" Bicycle Universe, May 2, 2003,http://bicycleaustin.info/buses-naturalgas.html (accessed February 12, 2004).

47. "Methane and Other Gases, Scientific Background and Other Research," U.S. Environmental Protection Agency, July 2003, http://www.epa.gov/ghginfo /topics/topic1.htm (accessed February 12, 2004).

48. "Fifteen Percent of Russia Badly Polluted, Says Putin," TerraDaily, June 4, 2003, http://www.edie.net/news/Archive/7078.cfm (accessed February 12, 2004).

49. "Drilling Called Big Methane Culprit: Southwest Oil and Gas Exploration Emits Much More Pollution Linked to Global Warming Than Once Thought, UC Irvine Scientists Find," October 7, 2003, http://www.latimes.com/news/nationworld /nation/la-na-oilsmog7oct07,1,6660933.story. (accessed October 8, 2003)

50. "Biggest U.S. Cities Don't Have Worst Smog: Study," October 6, 2003. http://www.cbc.ca/stories/2003/10/06/smog_samples031006 (accessed February 1, 2004).

51. "Glossary," Keyspan Energy Canada, http://www.keyspancanada.com /CUSTOMER/KEYSPAN/KEYSPAN.NSF/0/0AA655FEEFEDF97987256CF70 065D1AD?OpenDocument (accessed February 1, 2004).

52. "Sour Gas," Canadian Association of Petroleum Producers, http://www.capp .ca/default.asp?V_DOC_ID=767 (accessed February 1, 2004).

53. Andrew Nikiforuk, "Saboteurs, Sour Gas and You: Should You Be Concerned?" 2002, http://www.saboteursandbigoil.com (accessed February 1, 2004).

54. Editorial, "Oily Diplomacy," New York Times, August 19, 2002: "The case involves Exxon Mobil and its activities in the Indonesian province of Aceh. The Bush administration weighed in to discourage a lawsuit against the company filed on behalf of 11 Acehnese by the International Labor Rights Fund, a Washington group. The suit alleges that the company knew about and did nothing to stop murder, torture, and other crimes by security forces guarding its gas fields in Indonesia." http://www.laborrights.org/press/oilydiplomacy081902.htm (accessed February 12, 2004).

55. Adam Liptak, "Punitive Damages of $4.5 Billion in Exxon Valdez Case," New York Times, January 28, 2004, http://www.nytimes.com/2004/01/28/national /28CND-EXXON.html (accessed February 1, 2004).

56. The publication of The Limits to Growth: Thirty Year Update by Donella Meadows, Jorgen Randers, and Dennis Meadows in June 2004 (White River Junction, Vt.: Chelsea Green Publishing) may reignite the debate over unlimited economic and population growth.

Chapter 5: Gas Ability

1. Peta means 10^{15} or 1,000,000,000,000,000 or a million billion; tera means 10^{12} or 1,000,000,000,000 or a thousand billion. Peta is equivalent to a quadrillion. Tera is equivalent to a trillion, which is fortunate, and saves at least some confusion, since both begin with 't.'

2. J. H. Laherrère, A. Perrodon, and C. J. Campbell, "The World's Gas Potential" CD-ROM, Petroconsultants report, July, 1996, p. 200; and A. Perrodon, J. H. Laherrère, and C. J. Campbell, "The World's Non-Conventional Oil and Gas," *Petroleum Economist,* (March 1998), p. 113. Jean Laherrère confirms in private correspondence that these numbers are still relevant and accurate.

3. Tera and Peta have been used here, for explanatory purposes, but the U.S. industry, and therefore much of the world, uses Tcf (trillion cubic feet) and Bcf (billion cubic feet), and these will be used in the rest of this text. For 2001, Cedigaz reports 90.1 Tcf for marketed production, BP reports 89 Tcf for production excluding flaring and recycling, USDOE reports 90.7 Tcf for dry gas and OPEC 90.5 Tcf for marketed gas.

4. North Field (Qatar)—South Pars (Iran) is considered 5 to 6 times larger than the next largest field of Urengoy in the western Siberian basin.

5. Except for one brief year, discovery never picked up again. Information from "Report on 10-Year Trends (1993–2002) Shows Liquids Reserve Revisions and New Discoveries Have Outpaced Consumption," IHS Energy, October 2003, http://www.ihsenergy.com/company/press/pressreleases/arc2003/pr_100203 -10yrtrend.jsp (accessed February 7, 2004).

6. Much of the discovery is due to just the one field of North Field (Qatar)—South Pars (Iran). There is not enough gas infrastructure in either country to facilitate quick extraction of the reserve, and political problems will very likely also slow down production expansion.

7. The United States has an extraordinary combination of international trade debt and government deficit, of ballooning consumer debt, and of currency decline partly owing to this, and also to the desire of many nations to move to the euro and away from holding large amounts of a reserve currency associated with a hostile, dangerous, and unpredictable nation. Since in a loose way, the U.S. dollar is really only backed by its trillion-dollar debt to the world, and by oil, any large perturbation in foreign currency holdings or global oil supply could trigger a collapse a great deal worse than that in 1929.

8. Bernard Simon, "As Gold Surges, Many Mining Stocks Gain Favor," *New York Times,* January 25, 2004.

9. Richard Duncan, "The Olduvai Theory of Industrial Civilization," Institute on Energy and Man, December 1997, http://www.hubbertpeak.com/duncan /olduvai.htm (accessed February 7, 2004).

10. Jay Hanson, "Die Off," http://dieoff.org/ (accessed February 7, 2004).
11. See figure 14 in the section "Petroleum Overview and Crude Oil Production" of the "Annual Energy Review, 2002," U.S. Energy Information Administration, http://www.eia.doe.gov/emeu/aer/pdf/perspectives.pdf (accessed February 11, 2004).
12. It does not apply to a single well, either of gas or oil.
13. Though the case of the Ladyfern field in British Columbia shows that with enough wells, a large gas reserve can be emptied with remarkable speed. Andrew Nikiforuk, "Northern Greed," *Canadian Business,* May 12, 2003, http://www.canadianbusiness.com/features/article.jsp?content=20030512_53695_53695 (accessed February 9, 2004).
14. A province may be considered a collection of fields.
15. Gas also tends to leak away more easily from reservoirs over geological time.
16. For a discussion of the definitions of reserves, see chapter 2.
17. "Canada," Country Analysis Briefs, U.S. Energy Information Administration, January 2004, http://www.eia.doe.gov/emeu/cabs/canada.html (accessed February 11, 2004).
18. Table 1, "Natural Gas Annual 2002," U.S. Energy Information Administration, January 29, 2004, http://www.eia.doe.gov/pub/oil_gas/natural_gas/data_publications/natural_gas_annual/current/pdf/nga02.pdf (accessed February 11, 2004).
19. Official U.S. DOE reports, as of December 2003, still forecast U.S. gas production growth, yet the EIA report released on January 29, 2004, "Natural Gas Annual 2002" shows that the United States produced less gas in 2002 than in 2001. The report for 2003 production will not be available till the end of 2004 or early 2005. However, almost no independent analyst or even major production company thinks that gas production growth is possible, rather they confirm the decline indicated in the Natural Gas Annual 2002. See for instance, Brad Foss, "Natural Gas Markets Undergo Turbulent Transition As Production Declines," Associated Press, December 13, 2003, http://www.detnews.com/2003/business/0312/15/business-6813.htm (accessed February 11, 2004); and "Demand Outpaces Gas Production," *Houston Business Journal,* January 19, 2004, http://msnbc.msn.com/id/4001438/ (accessed February 11, 2004).
20. "Canada," Country Analysis Briefs, U.S. Energy Information Administration, January 2004, http://www.eia.doe.gov/emeu/cabs/canada.html (accessed February 11, 2004).
21. Outer Continental Shelf (OCS) includes the submerged lands, usually at depth of less than 600 feet, extending from the outer limit of the historic territorial sea (typically three miles), which is under state jurisdiction, to some undefined further outer limit. In the United States, the OCS is the portion of the shelf under federal jurisdiction. "Glossary," California Energy Commission, http://www.energy.ca.gov/glossary/glossary-o.html (accessed February 11, 2004).

22. For a map of CBM production, which constitutes an important part of unconventional gas, see figure 7 of "The Majors' Shift to Natural Gas," U.S. Energy Information Administration, September 2001, http://www.eia.doe.gov/emeu/finance/sptopics/majors/majors.pdf (accessed February 11, 2004).

23. For instance, see the "Executive Summary of Canadian Natural Gas: Review of 2002 & Outlook to 2015," Natural Resources Canada, November 2003, p. iv. http://www2.nrcan.gc.ca/es/erb/CMFiles/LandscapeFormat176OHC -18122003-2096.pdf (accessed February 7, 2004).

24. The highest estimates reach 160 trillion cubic feet.

25. "China's Message on Energy," *New York Times,* November 19, 2003, http://www.nytimes.com/2003/11/19/opinion/19WED2.html?ex=1069822800 &en=f9c8b3251e980a00&ei=5062 (accessed February 2, 2004).

26. The West Sole field is located in about 100 feet of water, at a depth of about 9,000 feet (2,745 meters). It is part of a similar geological structure as the giant Groningen onshore gas field in Holland, which was discovered in 1959. "History of Oil," BP, http://www.bp.com/genericarticle.do?categoryId=2010597&contentId=2015164 (accessed February 7, 2004).

27. The Cod Field was discovered in the British North Sea, and Ekofisk in the Norwegian area.

28. Leo Drollas and John Greenman, *Oil: The Devil's Gold* (London: Gerald Duckworth & Co., 1989), p. 203.

29. Carola Hoyos, "BP signs LNG Joint Venture with Algeria," *Financial Times,* October 25, 2003, http://www.bpamoco.org.uk/industry/03-10-25ft.htm (accessed February 7, 2004).

30. "Norway," Country Analysis Brief, U.S. Energy Information Administration, November 2003, http://www.eia.doe.gov/emeu/cabs/norway.html (accessed February 7, 2004).

31. The (oil and) gas-producing axis of the world is increasingly becoming an off-vertical swathe from northern Siberia through the Caspian Sea and Persian Gulf to northern and western Africa, which I have termed, with a nod in the direction of President Bush, the Axis of Gas.

32. The EIA gives Russia's reserves as 1,680 Tcf. The EIA is generally reliant on reserve numbers given by the countries concerned. Outside analysts differ, sometimes widely, in their reserve estimations, especially for Russia, as was discussed in chapter 2.

33. "The Russian Gas Industry to 2020," Research and Markets, http://www .researchandmarkets.com/reports/19819/19819.pdf (accessed February 2, 2004).

34. Petroleum Intelligence Weekly 42, no. 4 (January 26, 2004).

35. Reuters, "Rosneft, BP Do Not Find Oil on Two Sakhalin Blocks," December 26, 2003, http://uk.biz.yahoo.com/031226/80/ehv8r.html (accessed February 12, 2004).

36. "ExxonMobil and Chevron Texaco sacked from Sakhalin," *The Russian Journal,* January 30, 2004, http://www.russiajournal.ru/news/cnews-article.shtml?nd =42307 (accessed February 2, 2004).

37. "Yukos Battle Is Souring Interest in Russian Oil," *New York Times News Service,* Moscow, November 9, 2003, p. 11, http://www.taipeitimes.com/News/biz /archives/2003/11/09/2003075215 (accessed February 2, 2004).

38. Jim Carlton, "In Russia with Fragile Ecology: Stymied in Alaska, Oil Companies Find Russian Rules Aren't As Strict," *The Wall Street Journal,* September 4, 2002, http://www.sakhalin.environment.ru/en/sakhalin2/downloads/article_wst _carleton_09-02.htm (accessed February 2, 2004); "Sakhalin Energy Project a Threat to Russian Environment, Ecologists Say," Moscow (AFP), November 1, 2003, http://www.terradaily.com/2003/031101004534.966xkq5z.html (accessed February 2, 2004).

39. The Zapolyarnoe field, expected to reach an output of 700 Bcf per year in 2004. Wood Mackenzie, January 2004, personal communication

40. "The Russian Gas Industry to 2020," Research and Markets, http://www .researchandmarkets.com/reports/19819/19819.pdf (accessed February 2, 2004).

41. There is nothing "patriotic" about this: I am British, currently residing in Canada.

42. Venezuelan oil exports to the United States dropped dramatically after April 2002, and the attempt to remove President Hugo Chavez; but output has since risen, though the exact amount is disputed.

43. The United States imported 238 Bcf of LNG in 2001 according to the "Resource Development Trends and Emerging Issues," U.S. Energy Information Administration, January 24, 2003, http://www.eia.doe.gov/emeu/ perfpro/ch4sec9.html (accessed February 12, 2004).

44. The EIA gives Mexico gas reserves of 8.8 Tcf. "Mexico," Country Analysis Brief, U.S. Energy Information Administration, February 2003, http://www.eia.doe.gov/emeu/cabs/mexico.html (accessed February 12, 2004).

45. "Balancing Natural Gas Policy—Fueling the Demands of a Growing Economy (2003)," National Petroleum Council, http://npc.org/ (accessed February 7, 2004).

46. Mexico's population, 100 million in 2000, is expected to reach 114 million by 2010.

47. EIA: Venezuela has 148 Tcf in reserves, producing 1.1 Tcf in 2001, all of which is currently consumed domestically. "Venezuela," Country Analysis Brief, U.S. Energy Information Administration, May 2003, http://www.eia.doe.gov/emeu/cabs/venez.html (accessed February 11, 2004).

48. The EIA's figure is 4.4 tcf. "Colombia," Country Analysis Brief, U.S. Energy Information Administration, February 2003, http://www.eia.doe.gov/emeu/cabs /colombia.html (accessed February 7, 2004).

49. Ibid.

50. Colin Campbell, personal communication.

51. In 2002 they exported 68 percent of U.S. imports of LNG. The rest was Qatar at 16 percent, Algeria at 12 percent, and Nigeria at 4 percent. "Caribbean Fact Sheet," U.S. Energy Information Administration, June 2003, http://www.eia .doe.gov/emeu/cabs/carib.html (accessed February 3, 2004).

52. "U.S. Liquefied Natural Gas Imports from Trinidad (up to and including September 2003)," from"Natural Gas Inventory," U.S. Energy Information Administration, January 30, 2004. http://tonto.eia.doe.gov/dnav/ng/hist/ n9103td2M.htm (accessed February 3, 2004).

53. Trinidad and Tobago's reserve are estimated at 23.45 tcf by the EIA. "Trinidad & Tobago," Country Analysis Brief, U.S. Energy Information Administration, June 2003, http://www.eia.doe.gov/emeu/cabs/carib.html (accessed February 11, 2004).

54. "Professor Helps to Clear Hurdle for Trinidad and Tobago's Expansion," Alexander's Gas & Oil Connections, February 25, 2003, http://www.gasandoil .com/goc/company/cnl31209.htm (accessed February 3, 2004).

55. EIA "Caribbean Fact Sheet," U.S. Energy Information Administration, June 2003, http://www.eia.doe.gov/emeu/cabs/carib.html (accessed February 3, 2004).

56. This is higher than the EIA figure quoted above and reflects the fact reserve estimation relies greatly on judgment. See chapter 2 for a further discussion of reserve estimation.

57. Association for the Study of Peak Oil and Gas, ASPO Newsletter, no. 37 (January 2004), http://www.asponews.org/HTML/Newsletter37.html#302 (accessed February 3, 2004).

58. Daurius Figueira, "Globalized Gas: The Realities of Venezuela and Trinidad and Tobago," Petroleumworld, October 26, 2003, http://www.petroleumworld .com/SDY102603.htm (accessed February 3, 2004).

59. According to the EIA, Chile produced 40 Bcf in 2001, a fall of 44 percent since 1997. "Chile," Country Analysis Brief, U.S. Energy Information Administration, http://www.eia.doe.gov/emeu/cabs/chile.html (accessed February 11, 2004).

60. "Bolivia," Country Analysis Brief, U.S. Energy Information Administration, October 2003, http://www.eia.doe.gov/emeu/cabs/bolivia.html#gas (accessed February 8, 2004).

61. "Mexico," Country Analysis Brief, U.S. Energy Information Administration, October 2003, http://www.eia.doe.gov/emeu/cabs/mexico.html#gas (accessed February 8, 2004).

62. In January 2003 Peru had estimated reserves of 8.7 Tcf, but this does not include the Camisea field, estimated at 13 Tcf, which has not come into production yet.

"Peru," Country Analysis Brief, U.S. Energy Information Administration, April 2003, http://www.eia.doe.gov/emeu/cabs/peru.html (accessed February 11, 2004).

63. According to Oil and Gas Journal, as of January 2003, Argentina had 27 trillion cubic feet (Tcf). The Argentine government's estimate for natural gas reserves was slightly lower, at 23.4 Tcf. "Argentina," Country Analysis Briefs, U.S. Energy Information Administration, January 2004, http://www.eia.doe.gov/emeu/cabs /argentna.html (accessed February 11, 2004).

64. Brazil has also tried to import gas from Peru, but to date it has not come to pass.

65. The gas, in two fields, are 80 miles (130km) off the coast. "Santos Find Triples Brazil's Gas Reserves," E&P section, Petroleum Review, forthcoming.

66. It was 339 Bcf for 2001, according to the EIA. "Brazil," Country Analysis Briefs, U.S. Energy Information Administration, July 2003, http://www.eia.doe.gov/emeu/cabs/brazil.html (accessed February 11, 2004).

67. "Bolivia on the Brink of Civil War," New York Times, October 13, 2003, http://www.nytimes.com/2003/10/13/international/americas/13BOLI.html (accessed February 11, 2004).

68. The War of the Pacific, between Chile and Bolivia, took place between 1879 and 1883.

69. The EIA gives the following for reserves: Azerbaijan 30 Tcf; Kazakhstan 65 Tcf; Turkmenistan 71 Tcf; Uzbekistan 66 Tcf. "Caspian Sea Region" Country Analysis Briefs, U.S. Energy Information Administration, August 2003, http://www.eia.doe.gov/emeu/cabs/caspian.html (accessed February 8, 2004).

70. Wood MacKenzie gives the following for reserves: Azerbaijan 12 Tcf; Kazakhstan 54 Tcf; Turkmenistan 70 Tcf; Uzbekistan 38 Tcf.

71. The EIA gives the following production numbers for 2001: Azerbaijan 200 Bcf; Kazakhstan 360 Bcf; Turkmenistan 1.7 Tcf; Uzbekistan 2.23 Tcf. "Caspian Sea Region: Key Oil and Gas Statistics," U.S. Energy Information Administration, EIA statistical review, August 2003, http://www.eia.doe.gov/emeu/cabs/casp-stats.html (accessed February 8, 2004).

72. This is the Korpezhe-Kurt Kui pipeline. "Iran," Country Analysis Brief, U.S. Energy Information Administration, http://www.eia.doe.gov/emeu/cabs/iran-more.html (accessed February 12, 2004).

73. Assuming that the reserves as reported are reasonably accurate—an assumption that may not be justified.

74. Matt Simmons, "Is the Glass Half Full or Half Empty," May 26, 2003, http://www.simmonsco-intl.com/files/ASPO.pdf (accessed Feb 11, 2004). The area would roughly fit into California.

75. The EIA gives 812 Tcf. "Iran," Country Analysis Brief, U.S. Energy Information Administration, November 2003, http://www.eia.doe.gov/emeu/cabs/iran.html (accessed February 11, 2004), and the Oil & Gas Journal gives 940 Tcf.

76. "Iran," Country Analysis Brief, U.S. Energy Information Administration, November 2003, http://www.eia.doe.gov/emeu/cabs/iran.html (accessed February 11, 2004).

77. The Caspian littoral countries comprise Russia, Kazakhstan, Azerbaijan, Turkmenistan, and Iran.

78. http://www.eia.doe.gov/emeu/cabs/iran.html (accessed March 26, 2004).

79. The Bolivian "gas" uprising was discussed earlier in this chapter.

80. "Qatar," Country Analysis Brief, U.S. Energy Information Administration, November 2003, http://www.eia.doe.gov/emeu/cabs/qatar.html (accessed February 8, 2004).

81. "Saudi Arabia," Country Analysis Brief, U.S. Energy Information Administration, December 2003, http://www.eia.doe.gov/emeu/cabs/saudi.pdf (accessed February 13, 2004).

82. "Al-Naimi addresses LNG Summit in U.S.," in "Saudi Arabia," Country Analysis Brief, U.S. Energy Information Administration, December 2003, http://www.saudinf.com/main/y6504.htm (accessed February 11, 2004).

83. "Kuwait," Country Analysis Brief, U.S. Energy Information Administration," March 2003, http://www.eia.doe.gov/emeu/cabs/kuwait.html (accessed February 11, 2004).

84. Ibid.

85. "Iraq," Country Analysis Brief, U.S. Energy Information Administration, http://www.eia.doe.gov/emeu/cabs/iraq.html (accessed February 11, 2004).

86. Nelson Antosh, "Ex-Oil Minister Warns Iraq Fields Being Ruined," *Houston Chronicle,* February 10, 2004, http://www.chron.com/cs/CDA/ssistory.mpl /business/energy/2397157 (accessed February 11, 2004).

87. The Gulf of Guinea, for the purposes of hydrocarbon extraction, includes Angola, Benin, Cameroon, Congo, Brazzaville, Côte d'Ivoire, Equatorial Guinea, Gabon, Ghana, Nigeria, São Tomé and Príncipe.

88. Algeria is reported to have gas reserves ranging from 160 Tcf to 204 Tcf. "Algeria," Country Analysis Brief, U.S. Energy Information Administration, January 2003, http://www.eia.doe.gov/emeu/cabs/algeria.html (accessed February 13, 2004).

89. Ibid.; Algeria began producing gas in 1961 and exporting LNG in 1964.

90. Reported Libyan reserves are 46 Tcf with a potential of perhaps 50 to 70 Tcf. "Libya," Country Analysis Brief, U.S. Energy Information Administration, January 2004, http://www.eia.doe.gov/emeu/cabs/libya.html (accessed February 11, 2004).

91. Egypt's reserves are now set at 58.5 Tcf, but may be as high as 120 Tcf. "Egypt," Country Analysis Brief, U.S. Energy Information Administration, January 2003, http://www.eia.doe.gov/emeu/cabs/egypt.html (accessed February 11, 2004).

92. Indonesia appears to have peaked in 1999 at 3 Tcf. "Indonesian 2001 Natural Gas Developments," U.S. Embassy, Indonesia, 2001, http://www.usembassy-jakarta.org/econ/natural_gas2001.html (accessed February 11, 2004).

Chapter 6: Where on Earth Are We Now?

1. Unfortunately there is no generally accepted agreement for the meaning of "conventional" oil, and in addition, useable oil is produced by the refining process and from gas production.

2. Jamie Miyazaki, "Natural Gas's New Global Role," *Asia Times Online,* November 11, 2003, http://www.atimes.com/atimes/Global_Economy/EK11Dj01.html (accessed February 13, 2004).

3. Theodore Roosevelt told his attorney general privately that Standard Oil's directors were "the biggest criminals in the country." Quoted in Daniel Yergin, *The Prize: The Epic Quest For Oil, Money & Power* (New York: Simon and Schuster, 1991), p. 108.

4. Ibid., p. 110.

5. The year 1720 saw the bursting of two of the earliest great stock market bubbles, The South Sea Bubble and the Mississippi Bubble, though the first, and still perhaps most famous, bubble was Tulipomania, which reached its dizzying heights in the Netherlands in 1630s. See Charles MacKay, *Extraordinary Popular Delusions and the Madness Of Crowds* (1841), http://www.litrix.com/madraven /madne001.htm (accessed February 13, 2004).

6. "U.K. Trade in Crude Oil and Petroleum Products," U.K. Department of Trade and Industry, August 2000, http://www.dti.gov.uk/energy/inform/energy _trends/articles/bpaug2000.pdf (accessed February 13, 2004).

7. Thatcher blamed the miners and unions for the fall of Edward Heath's Conservative government in February 1974.

8. See Dan Berman & John O'Connor, *Who Owns the Sun?* (White River Junction, Vt.: Chelsea Green Publishing, 1996), chap. 4, "Public Power," and also Eugene Coyle's remarks on monopolies in the same volume.

9. "Balancing Natural Gas Policy—Fueling the Demands of a Growing Economy," National Petroleum Council, September 2003, http://www.npc.org/reports /ng.html (accessed February 4, 2004).

10. George W. Bush, son of previous president, George H. W. Bush.

11. Canada also has a winter gas storage system that is filled during the warmer months.

12. Matt Simmons, the well-known energy banker in Houston mentioned previously, has estimated that up to forty LNG terminals may be needed by 2010 alone. See "The Natural Gas Riddle: Why Are Prices So High? Is a Serious Crisis Underway?" Simmons and Company International, December 11, 2003, http://www.simmonsco-intl.com/files/IAEE%20Mini%20Conf.pdf (accessed February 4, 2004).

13. Associated Press, "Wyoming's Run of Natural Gas Growth Could End," January 29, 2004, http://www.casperstartribune.net/articles/2004/01/29/news/wyoming /6273e36d8a9c63c487256e2a00812b08.txt (accessed February 4, 2004).

14. At over 15,000 feet, the wells can cost $20 million each.

15. "Secretary Norton Unveils New Incentives to Boost Domestic Natural Gas Production, Save Americans $570 Million a Year," U.S. Department of Interior, January 23, 2004, http://www.doi.gov/news/040123a.htm (accessed February 4, 2004).

16. The North American Free Trade Agreement, signed in 1992 by Canada, Mexico, and the United States, took effect in 1994.

17. This figure includes Mexico, the United States, and Canada.

18. Because of the gas price rises and the possibility of an actual gas shortage, there is continuing official discussion about using nuclear reactors to supply the energy needed for extraction and production.

19. Pemex is Mexico's national oil and natural gas company.

20. Anthony Harrup, "Texas Looks to Mexico for Future Natural Gas Supplies," *Dow Jones Newswires,* August 25, 2003, http://www.latinpetroleum.com /article_1782.shtml (accessed February 13, 2004).

21. Britain is discussed separately earlier in this chapter.

22. The Axis of Gas, as I have termed it, runs from Siberian Russia, through the Caspian basin, the Persian Gulf, to North and West Africa. It contains the great majority of the world's remaining natural gas.

23. See http://www.eurogas.org/index2.htm (accessed February 13, 2004).

24. In 2001 China extracted and used about 1 Tcf of natural gas.

25. At the beginning of 2003, China's gas reserves were 53.3 tcf according to the EIA. "China," Country Analysis Briefs, U.S. Energy Information Administration, June 2003, http://www.eia.doe.gov/emeu/cabs/china.html (accessed February 13, 2004).

Chapter 7: Energy Security

1. The Last Post is a traditional bugle call played in military forces at the end of the day. It is also played at military and other funerals.

2. Alan Reynolds, "Gas Lines for California?" Town Hall.com, September 4, 2003, http://www.townhall.com/columnists/alanreynolds/ar20030904.shtml (accessed February 13, 2004).

3. "President Focuses on Energy Security in Radio Address," White House, February 23, 2002, http://www.whitehouse.gov/news/releases/2002/02/20020223.html (accessed February 13, 2004).

4. "The Rise of Japan 1853–1914," Imperial War Museum, http://www.iwm.org .uk/online/pearl_harbour/hb_rise_of_japan.htm (accessed February 13, 2004).

5. See note 22 in chapter 6. For a full explanation of the Axis of Gas see p. 111–12, in chapter 5, Gas Ability.

6. This is a catchy misquote of John Muir's original "When we try to pick out anything by itself, we find it hitched to everything else in the Universe." It is from Muir's *My First Summer in the Sierra* (Boston: Houghton Mifflin, 1911), chap. 6.

7. There are other organizations such as Wood Mackenzie and Cedigaz, who work hard to achieve accurate numbers, but they sometimes produce results that diverge considerably, a testament to the difficulties involved.

8. As explained in chapter 3, most of Canada's remaining oil resource is really bitumen, and, it is hard, expensive, and environmentally damaging to mine and make into synthetic crude. Thus the constraints on Canadian oil production are financial and technological.

9. These estimates are based on 2001 figures from the EIA. The situation can only get worse, unless the experiments on methane hydrates in the Nankai Trough prove successful with both extraordinary and unlikely speed.

10. This is technically true at present but a little unfair! Uranium is quite widespread in the world, including in seawater, which Japan has plenty of, but that does not mean it is economically recoverable. Japan has no economically significant deposits and mines no uranium of its own. "Nuclear Power in Japan," World Nuclear Association, January 2004, http://www.world-nuclear.org/info/inf75.htm (accessed February 13, 2004); and Energy Study Nuclear Fuels, German Federal Institute for Geosciences and Natural Resources, http://www.bgr.de/index.html?/b123/e_esuran.html (accessed February 13, 2004).

11. Japan has ships as part of its Self-Defense Force, but they are not used to protect its energy supplies.

12. Richard Duncan, *The Dollar Crisis: Causes, Consequences, Cures* (Singapore: J. Wiley & Sons (Asia), 2003.), p. viii.

13. "Japan Calls on U.S. to Reduce Bases," *BBC,* November 16, 2003, http://news .bbc.co.uk/2/hi/asia-pacific/3274877.stm (accessed February 13, 2004). The pressure for U.S. withdrawal was dramatically increased after U.S. soldiers raped a Japanese schoolgirl in 1995 on Okinawa Island.

14. Brunei, China, Indonesia, Malaysia, the Philippines, Taiwan, and Vietnam.

15. The Spratly Islands consist of over 100 islands or reefs but cover less than three square miles.

16. Michael T. Klare, *Resource Wars* (New York: Metropolitan Books, 2001).

17. More than half of the world's merchant fleet sails through the South China Sea. "South China Sea Region," Country Analysis Briefs, U.S. Energy and Information Administration, September 2003, http://www.eia.doe.gov/ emeu/cabs/schina.html (accessed February 4, 2004).

18. Lester R. Brown, *Who Will Feed China? Wake-Up Call for a Small Planet,* (Washington, D.C.: Worldwatch Institute, 1995).

19. "Evolution of the World Grain Production, Comparison with China and United States," Le Monde diplomatique, http://carto.eu.org/article2474.html (accessed February 13, 2004).

20. That is, if all goes according to plan. "Three Gorges Dam Project," China Online, October 3, 2003, http://www.chinaonline.com/refer/ministry_profiles /threegorgesdam.asp (accessed February 13, 2004).

21. Though some have suggested that the tar sands operation will be constrained first by pollution or water problems. Nuclear power has been suggested as a substitute for natural gas energy. The bitumen itself can also be used as an energy source. All these possibilities are more expensive and troublesome than using gas. See chapter 3 for more discussion of tar sands.

Chapter 8: Where the Hell Are We Going?

1. UK Department of Trade and Industry, Energy White Paper, "Our energy future - creating a low carbon economy" February 2003, http://www.dti .gov.uk/energy/whitepaper/ourenergyfuture.pdf (accessed May 25, 2004).

2. The "vert" is pronounced as the French word for "green," not as in "extrovert." Thus it rhymes with "hair" not "hurt."

3. Namely, by Matt Simmons, president of the world's largest private energy bank, and who was an adviser on the National Petroleum Council and was on the Bush Energy Transition Team in 2001.

4. "Secretary Norton Unveils New Incentives to Boost Domestic Natural Gas Production, Save Americans $570 Million a Year," U.S. Department of the Interior, January 23, 2004, http://www.doi.gov/news/040123a.htm (accessed February 4, 2004).

5. U.S. House Subcommittee on Energy and Mineral Resources, 2002 (testimony of Diemer True, chairman, Independent Petroleum Association of America).

6. "Alaska Natural Gas Subsidy Dies: Ottawa Wins Rollbacks on Floor Price, Sensitive Drilling," National Post, November 15, 2003.

7. Ibid., quoting Bill Tauzin, chairman of the House Energy and Commerce Committee.

8. Governor Jeb Bush of Florida. National Governors Association, http://www.nga .org/governors/1,1169,C_GOVERNOR_INFO%5ED_124,00.html (accessed February 13, 2004).

9. Governor Arnold A. Schwarzenegger of California. National Governors Association, http://www.nga.org/governors/1,1169,C_GOVERNOR_INFO^D_1051,00.html (accessed February 13, 2004).

10. The ethanol must be triple-distilled, using a great deal of energy, and much, if not all, of the corn is produced using standard oil- and gas-intensive, industrial agriculture. The corn is quite likely to be genetically modified, which though of less concern from an energy perspective, is of the greatest concern when viewed from the deteriorating biosphere.

11. "Pork" or "pork-barrel politics" refers to the state or national treasury, into which politicians and government officials dip for "pork," or funds for local projects. The phrase is probably derived from the pre-Civil War practice of periodically distributing salt pork to the slaves from huge barrels. William Safire, *Safire's New Political Dictionary* (New York: Random House, 1993), http://phrases.shu.ac.uk /bulletin_board/24/messages/1197.html (accessed Februrary 13, 2004).

12. Note that it is sheer quantity of all cars, which is the problem, not just SUVs.

13. *U.S. Energy Policy Act of 2003* (November 17, 2003), Title VII, Section 774.

Chapter 9: But What Else Can We Do?

1. Jimmy Carter, "The President's Proposed Energy Policy," April 18, 1977, PBS, http://www.pbs.org/wgbh/amex/carter/filmmore/ps_energy.html (accessed February 10, 2004).

2. "How It All Began," Earth Day Network, http://www.earthday.net/about/history.stm (accessed February 6, 2004).

3. Ibid.

4. "Monthly Energy Review, January 2004," U.S. Energy Information Administration, http://www.eia.doe.gov/emeu/mer/pdf/pages/sec1_17.pdf (accessed February 6, 2004).

5. Steven Bernstein, "Liberal Environmentalism and Global Environmental Governance," *Global Environmental Politics* 2, no. 3 (August 2002).

6. See note 78 in chapter 2 for a description of "neoliberal."

7. Donella H. Meadows, Jorgen Randers, and Dennis Meadows, *The Limits to Growth: A Report for the Club of Rome's Project on the Predicament of Mankind* (New York: Universe Books, 1972).

8. Quoted in Steven Bernstein, "Liberal Environmentalism and Global Environmental Governance," with reference to *Limits to Growth*.

9. C. J. Campbell, ed., *The Essence of Oil and Gas Depletion* (Brentwood: Multi-Science Publishing UK, 2003).

10. The exact year of global oil peak will be hard to pinpoint until sometime afterward, and even then, there is no consensus about what constitutes conventional oil, which means that if one includes "all liquids"—polar, deep-sea, heavy oil, and more—the peak will come later, but perhaps not much later. It is safer to say that the 2000–10 decade will be a bumpy plateau for oil production; the year 2004 got off to a very rocky start indeed.

11. I have termed this the "cyanide solution," since coal + nuclear, or CN, is the chemical formula of cyanide.

12. "The French Fast-Breeder Programme," http://www.wise-paris.org/english /ournewsletter/1/page6.html (accessed February 12, 2004); Paul Brown,

"Decaying and Dangerous, the Legacy of a Flawed Nuclear Vision," *The Guardian,* August 26, 2003, http://politics.guardian.co.uk/green/story/0,9061 ,1029405,00.html (accessed June 17, 2004).

13. "Earthquakes in the Vicinity of Yucca Mountain," http://www.state.nv.us /nucwaste/yucca/seismo01.htm (accessed February 12, 2004); "An Evaluation of Key Elements in the U. S. Department of Energy's Proposed System for Isolating and Containing Radioactive Waste," U.S. Nuclear Waste Technical Review Board, November, 2003, http://www.nwtrb.gov/reports/mlc019.pdf (accessed February 12, 2004).

14. "It's impossible. And what's more, it's improbable," *The Economist,* July 18, 2002.

15. A look at the evidence, however, shows that there are great differences between the case for oil peak, and that for fusion. Without a miracle or maybe two, oil peak will happen long before any fusion reactor reaches ignition—the point when it generates more power out than it took to get started.

16. Hugh Sharman, "The Dash for Wind: West Denmark's Experience and U.K.'s Energy Aspirations," Incoteco (Denmark) ApS, May 2003, http://www.glebe-mountaingroup.org/Articles/DanishLessons.pdf (assessed February 4, 2004).

17. Richard Heinberg, *Powerdown: Options and Actions for a Post-Carbon World* (Gabriola Island, BC: New Society, 2004), and Meadows et al., *The Limits to Growth: 30-Year Update* (White River Junction, Vt.: Chelsea Green Publishing, 2004), are two of the rather small number of books that openly consider contraction of both population and economy, which the evidence now suggests is the only serious policy path for the rest of this century. My forthcoming book, "Replacing the Future: Global Relocalization—Learning to Live in a Limited World," will as the title suggest, explore the building of the kind of infrastructure and institutions that low-energy, post-carbon societies will require.

18. World Commission on Environment and Development (WCED), *Our Common Future* (Oxford: Oxford University Press, 1987).

19. http://www.un.org/News/Press/docs/2003/sgsm8974.doc.htm (accessed March 25, 2004).

20. Emmanuel Yashim, "Third World Loses $200bn through Capital Flight—U.N. Secretary General Kofi Annan," *Daily Trust* (Abuja), November 5, 2003, original site: http://allafrica.com/stories/200311050296.html, also available at http://www .tagstudio.net/mumbai/mt/archives/000047.html (accessed February 6, 2004).

21. Gro Harlem Brundtland, "Health & Population," *BBC,* 2000, http://news.bbc.co.uk/hi/english/static/events/reith_2000/lecture4.stm (accessed February 13, 2004).

22. Petroleum geologist Colin Campbell's euphemism for contraction. Association for the Study of Peak Oil and Gas, ASPO Newsletter no. 38 (February 2004) http://www.asponews.org/ (accessed February 12, 2004).

23. Strictly speaking, it is the "balance of payments" debt.

24. In answer to the question "Who owns the Federal Reserve?" the Federal Reserve FAQ says the following: "The Federal Reserve System is not 'owned' by anyone and is not a private, profit-making institution. Instead, it is an independent entity within the government, having both public purposes and private aspects." Others claim that it is quite simply a private bank masquerading as a public entity. To say the least, the answer given above is obfuscatory, perhaps deliberately so. http://www.federalreserve.gov/faq.htm (accessed March 25, 2004).

25. One of the best books on local money systems is Richard Douthwaite, *Short Circuit: Strengthening Local Economics for Security in an Unstable World* (Dublin, Ireland: Lilliput Press, 1996).

26. In other words using no oil or natural gas in part of the process of producing food.

27. If the reader doubts this is possible, local history books are full of community endeavors that worked for generations. Particularly interesting examples, which demonstrate how traditional techniques actually work, are books by John Seymour such as *The Self-Sufficient Life and How to Live It: The Complete Back-to-Basics Guide* (New York: DK Publishing, 2003; First published 1976).

28. If indeed they didn't start that way.

29. By "post-carbon" I mean not using hydrocarbons, especially nonrenewable hydrocarbons. In reality, thanks to the extraordinary properties of hydrocarbons, we shall continue to use some, but they should at least be from renewable sources.

30. Joel Bakan, *The Corporation: The Pathological Pursuit of Profit and Power* (New York: Free Press 2004).

31. Aurora Institute, http://www.aurora.ca/info/structure.htm (accessed February 10, 2004).

32. Bakan, *The Corporation,* 57.

33. Bakan, *The Corporation,* 158.

34. John Kenneth Galbraith's *The Great Crash, 1929* (Boston: Houghton Mifflin 1955) is a readable and masterly account.

35. See for instance the Program on Corporations, Law & Democracy http://www.poclad.org (accessed February 12, 2004).

36. Population expert Virginia Abernethy recognizes this in her paper "Fossil Fuel Energy and Fertility Rates," AAAS, January 2004.

37. Clive Ponting, "The Lessons of Easter Island," http://www.eco-action.org/dt/eisland.html (accessed February 13, 2004).

38. Ken Livingstone, mayor of London, publicly advocates that world population needs to be reduced by "perhaps half of what it now is over a century." *The Ecologist* 33, no. 10 (December 2003/January 2004): 21.

39. China "advocates" delayed marriage and delayed child bearing, fewer and healthier births, and one child per couple.

40. Bill McKibben, *Maybe One* (New York: Simon & Schuster, 1998).

41. Some have obviously been breached—extinction is one marker of this.

42. In fairness, the Sierra Club, a major orthodox green group, has been discussing population growth, but at this writing, there is a bitter battle going on for the leadership of the club because of this issue. See Steven Rosenfeld, "Population Bombshell," http://www.tompaine.com/feature2.cfm/ID/9913 (accessed February 11, 2004).

43. Jimmy Carter, State of the Union Address, 1980, http://www.jimmycarterlibrary.org/documents/speeches/su80jec.phtml (accessed February 10, 2004).

44. Sheep fleece insulation and light-clay straw are two possibilities.

45. "Why Insulate Your Home?" Lancashire Energy Efficency Advice Center, http://www.leeac.org.uk/Insulation/home_insulation.html (accessed February 10, 2004).

46. Some buildings are superinsulated and even in cold climates need very little extra heating.

47. http://www.chelseagreen.com/item_detail.php?id=595 (accessed February 10, 2004).

48. Available in print and online at http://www.homepower.com (accessed February 10, 2004).

49. There are many books and manuals about constructing buildings that use less energy–the Internet will help in finding something locally appropriate. Two examples of respected books are William M. C. Lam, *Sunlighting as Formgiver for Architecture* (New York: Van Nostrand Reinhold, 1986) and James Kachadorian, *The Passive Solar House* (Real Goods Independent Living Books) (White River Junction, Vt.: Chelsea Green Publishing Company, 1997.)

50. Irish Department of Environment, Heritage and Local Government, May 2003, http://www.environ.ie/DOEI/DOEIPub.nsf/enSearchView/3108E941225E013 180256D3400308D41?OpenDocument&Lang=en, (accessed February 10, 2004).

51. Needless to say this book is being typed into a computer. Furthermore, I run an Internet broadcasting station, GlobalPublicMedia.com, which though modest in its needs, nonetheless requires at least one or two computers running all the time.

52. Richard Heinberg's term from his book of the same name, *Powerdown: Options and Actions for a Post-Carbon World* (Gabriola Island: New Society, 2004).

53. Albert A. Bartlett, "Forgotten Fundamentals of the Energy Crisis," http://www .npg.org/specialreports/bartlett_section3.htm (accessed February 8, 2004).

54. There are some techniques in the research stage that may allow industrial production of nitrogen fertilizer without using natural gas, but that will do nothing to address the problem that we are using too much nitrogen fertilizer for the soil.

55. On the occasion flesh is eaten, it should at least be locally produced, again with no oil or gas intervention.

56. Thanks to determined efforts by well-organized, local citizens.

57. There have been numerous independent studies that heavily criticize the hydrogen economy, but in January 2004, to the amazement of many, a U.S. government-funded study came out with very similar criticisms in *The Hydrogen Economy: Opportunities, Costs, Barriers, and R&D Needs* by the National Academy of Engineering (NAE) and Board on Energy and Environmental Systems (BEES), published by the National Academies Press, http://books.nap.edu/books/0309091632/html/index.html (accessed February 13, 2004).

58. For more information on local money-energy-food systems, please contact the Post-Carbon Institute, http://www.postcarbon.org (accessed June 17, 2004).

59. See chapter 4, footnote 3 for the laws of thermodynamics.

60. A post-carbon world would also see the kinds of dramatic reductions in CO_2 emissions, which many believe are required to reduce climate change.

61. With the possible exception of a few deep-sea, volcanic-vent, sulfur-based creatures.

62. Not least because they have been driven out of business.

63. "Eco-footprinting poses a single, simple question: How much productive land and water (i.e., ecosystem area) is required to support the study population at a specified material standard of living?" William Rees, Impeding Sustainability? The Ecological Footprint of Higher Education, Planning for Higher Education March–May 2003.

64. "Vancouver Co-operative Auto Network," http://www.cooperativeauto.net/ and Car Share Network, an international list of car-sharing organizations http://www.carsharing.net/index.html (accessed February 10, 2004).

65. "Vancouver Promotes Car-Sharing," http://www.mirabilis.ca/archives/001277.html (accessed June 17, 2004).

66. Which is increasingly imported from Trinidad.

67. What I do advocate are measures that conserve and use energy more efficiently, but as part of a program of full-scale contraction.

Selected Bibliography and Further Reading

The following books are mainly references for chapter 9. Most of the references for the discussion of natural gas itself are from hundreds of articles, which are referenced in the notes. Special mention must be given to the work of Jean Laherrère, Walter Youngquist, and Colin Campbell (books mentioned below), references to whom can also be found in the list of Online Resources.

Abernethy, Virginia. *Population Politics: The Choices That Shape Our Future.* New York: Insight Books, 1993.

Bakan, Joel. *The Corporation: The Pathological Pursuit of Profit and Power.* New York: Free Press, 2004.

Berman, Daniel M., and John T. O'Connor. *Who Owns the Sun?: People, Politics, and the Struggle for a Solar Economy.* White River Junction, Vt.: Chelsea Green Publishing Company, 1996.

Bernstein, Steven. "Liberal Environmentalism and Global Environmental Governance," *Global Environmental Politics* 2, no. 3 (August 2002).

Campbell, Colin. *The Coming Oil Crisis.* Brentwood, U.K.: Multi-Science Publishing Company & Petroconsultants SA, 2004.

————. *The Essence of Oil and Gas Depletion: Collected Papers and Excerpts.* Brentwood, U.K.: Multi-Science Publishing Company, 2003.

Catton, William Robert. *Overshoot: The Ecological Basis of Revolutionary Change,* Urbana, Ill.: University of Illinois Press, 1980.

Deffeyes, Kenneth S. *Hubbert's Peak: The Impending World Oil Shortage.* Princeton: Princeton University Press, 2001.

Douthwaite, Richard. J. *Short Circuit: Strengthening Local Economies for Security in an Unstable World.* Dublin, Ireland: Lilliput Press, 1996.

————. *The Growth Illusion: How Economic Growth Has Enriched the Few,*

Impoverished the Many and Endangered the Planet. Gabriola Island, B.C.: New Society Publishers, 1999.

Drollas, Leo, and Jon Greenman. *Oil: The Devil's Gold,* London: Duckworth, 1989.

Duncan, Richard. *The Dollar Crisis: Causes, Consequences, Cures.* London: John Wiley & Sons, 2003.

Ehrlich, Paul R., and Anne H. Ehrlich. *The Population Explosion.* New York: Simon and Schuster, 1990.

Ewen, Stuart. *PR!: A Social History of Spin.* New York: Basic Books, 1996.

Gever, John, et al. *Beyond Oil: The Threat to Food and Fuel in the Coming Decades.* Cambridge, Mass: Ballinger Pub. Co., 1986.

Hardin, Garrett James. *The Ostrich Factor: Our Population Myopia.* New York: Oxford University Press, 1999.

Heinberg, Richard. *The Party's Over: Oil, War, and the Fate of Industrial Societies.* Gabriola Island, B.C.: New Society Publishers, 2003.

——————. *Powerdown: Option and Actions for a Post Carbon World.* Gabriola Island, B.C.: New Society Publishers, 2003.

Johnson, Chalmers A. *Blowback: The Costs and Consequences of American Empire.* New York: Metropolitan Books, 2000.

Kelsey, Jane. *Economic Fundamentalism.* London and East Haven, Conn.: Pluto Press, 1995.

Klare, Michael T. *Resource Wars: The New Landscape of Global Conflict.* New York: Metropolitan Books, 2001.

Korten, David C. *When Corporations Rule the World.* San Francisco: Berrett-Koehler Publishers, and Bloomfield, Conn.: Kumarian Press, 2001.

Kunstler, James Howard. *The Geography of Nowhere: The Rise and Decline of America's Man-Made Landscape.* New York: Simon and Schuster, 1993.

——————. *The Long Emergency.* New York: Grove-Atlantic, 2004.

Lens, Sidney. *The Forging of the American Empire.* New York: Crowell, 1971.

Loewen, James W. *Lies My Teacher Told Me: Everything Your American History Textbook Got Wrong.* New York: New Press, 1995.

McKibben, Bill. *Maybe One: a Personal and Environmental Argument for Single-Child Families.* New York: Simon and Schuster, 1998.

Meadows, Dennis, Jorgen Randers, and Donella Meadows. *The Limits to Growth: The 30-Year Global Update.* White River Junction, Vt.: Chelsea Green Publishing Company, 2004.

Merkel, Jim. *Radical Simplicity: Small Footprints on a Finite Earth.* Gabriola Island, B.C.: New Society Publishers, 2003.

Pahl, Greg. *Natural Home Heating: The Complete Guide to Renewable Energy Options.* White River Junction, Vt.: Chelsea Green Publishing Company, 2003.

Pimentel, David, and Marcia Pimentel. *Food, Energy, and Society.* Niwot, Colo.: University Press of Colorado, 1996.

Rowbotham, Michael. *The Grip of Death: A Study of Modern Money, Debt Slavery and Destructive Economics.* Charlbury, Oxfordshire: Jon Carpenter, 1998.

Schmookler, Andrew Bard. *The Illusion of Choice: How the Market Economy Shapes Our Destiny.* Albany: State University of New York Press, 1993.

Seabrook, Jeremy. *The Myth of the Market: Promises and Illusions.* London: Green Books, 1990.

Shrybman, Steven. *The World Trade Organization: A Citizen's Guide.* Toronto: James Lorimer, 2001.

Tainter, Joseph A. *The Collapse of Complex Societies.* New York: Cambridge University Press, 1988.

Trainer, Ted. *Developed to Death: Rethinking Third World Development.* London: Green Print, 1989.

Yergin, Daniel. *The Prize: The Epic Quest for Oil, Money, and Power.* New York: Simon and Schuster, 1991.

Youngquist, Walter. *Geodestinies: The Inevitable Control of Earth Resources over Nations and Individuals.* Portland, Or: National Book Co., 1997.

INDEX

Books for Sustainable Living

Chelsea Green Publishing
PO Box 428
White River Junction VT 05001-0428

Use this card as a bookmark, then tell us what you think . . .

Chelsea Green publishes books on a variety of subjects related to sustainable living. Return this card for a complete catalog of our books. Please indicate the subject(s) that interest you most:

_Renewable Energy __Shelter __Food __Gardening __Nature __Environment

Other topics that interest you: _____

What publications do you read regularly? _____

In which book was this inserted? _____ Where purchased? _____

How would you describe your satisfaction with this book?

_Exceeded expectations __Met expectations __Could be improved __Disappointed

Comments for the author or publisher: _____
_____
_____

Name: _____

Address: _____ E-mail: _____

☐ I'd like to receive updates and special offers from Chelsea Green.

www.chelseagreen.com

Thank you very much!

CHELSEA GREEN PUBLISHING CO.